ROCK AND TEMPEST

Patricia Collins was a Wran (the women's branch of the RAN) stationed in Darwin at the time of Cyclone Tracy. She wrote a brief account of her experiences during Tracy and the clean-up for her son when he was a teenager and interested in what had happened. The essay was submitted by a third party to the editor of the magazine *Australian Warship* and was subsequently published with her permission in Issue 24 in 2005. Patricia then turned it into a book and brought in experiences from many of those she served with during that time. *Rock and Tempest* is her first book.

ROCK AND TEMPEST

SURVIVING CYCLONE TRACY AND ITS AFTERMATH

PATRICIA COLLINS

hachette
AUSTRALIA

hachette
AUSTRALIA

Published in Australia and New Zealand in 2024
by Hachette Australia
(an imprint of Hachette Australia Pty Limited)
Gadigal Country, Level 17, 207 Kent Street, Sydney, NSW 2000
www.hachette.com.au

Hachette Australia acknowledges and pays our respects to the past, present and future Traditional Owners and Custodians of Country throughout Australia and recognises the continuation of cultural, spiritual and educational practices of Aboriginal and Torres Strait Islander peoples. Our head office is located on the lands of the Gadigal people of the Eora Nation.

NATIONAL LIBRARY OF AUSTRALIA

A catalogue record for this book is available from the National Library of Australia

ISBN: 978 0 7336 5023 9 (paperback)

Cover design by Luke Causby/Blue Cork
Front cover photograph courtesy of Bruce Howard/Newspix
Author photograph courtesy of Lee-Ann Wilson Photography
Typeset in Bembo Std by Kirby Jones
Printed and bound in Australia by McPherson's Printing Group

MIX
Paper | Supporting responsible forestry
FSC www.fsc.org FSC® C001695

The paper this book is printed on is certified against the Forest Stewardship Council® Standards. McPherson's Printing Group holds FSC® chain of custody certification SA-COC-005379. FSC® promotes environmentally responsible, socially beneficial and economically viable management of the world's forests.

To all those who served in the WRANS,
especially Lyn Baker (nee Alderton),
Joan Smolinski (nee Czarnecki)
and the late Ruth Clarke (nee Turnbull)

THE NAVAL HYMN

Eternal Father, strong to save,
Whose arm hath bound the restless wave,
Who bidd'st the mighty ocean deep
Its own appointed limits keep;
Oh, hear us when we cry to Thee,
For those in peril on the sea!

O Christ! Whose voice the waters heard
And hushed their raging at Thy word,
Who walked'st on the foaming deep,
And calm amidst its rage didst sleep;
Oh, hear us when we cry to Thee,
For those in peril on the sea!

Most Holy Spirit! Who didst brood
Upon the chaos dark and rude,
And bid its angry tumult cease,
And give, for wild confusion, peace;
Oh, hear us when we cry to Thee,
For those in peril on the sea!

O Trinity of love and power!
Our brethren shield in danger's hour;
From rock and tempest, fire and foe,
Protect them wheresoe'er they go;
Thus evermore shall rise to Thee
Glad hymns of praise from land and sea.

William Whiting, 1860

CONTENTS

INTRODUCTION

On Christmas Eve 1974, Cyclone Tracy hit Darwin in Australia's Northern Territory. By dawn on Christmas Day, at least fifty people were dead, seventeen lost at sea, hundreds injured and approximately 45,000 people were homeless.

I was one of about ninety Wrans based at HMAS *Coonawarra*, a Royal Australian Navy shore base on the outskirts of Darwin. The cyclone severely damaged our infrastructure and destroyed our headquarters at HMAS *Melville*. One Navy patrol boat was sunk with the loss of two crewmen and another patrol boat beached on rocks. One sailor lost his wife and two children, and another lost his young son. Most of us were left with only the clothes we were wearing and had no electricity, running water or sanitation.

My friends and I had started our service in the Women's Royal Australian Naval Service (WRANS) with enthusiasm and youthful high spirits. We had no idea that in less than two years, we would endure one of the most devastating natural disasters in our country's history and take part in Operation Navy Help Darwin, the Royal Australian Navy's greatest effort to relieve suffering caused by a weather event.

In the first few months after Cyclone Tracy devastated Darwin, those of us who'd been there were asked, 'What was it like?' The sailors who travelled up to assist us in

Operation Navy Help Darwin returned home to be asked, 'What was it like?'

At first, many of us tried to explain but people couldn't understand. We were met with incomprehension and often disbelief. 'You had to be there,' became a standard fallback. 'It' was too hard to describe. 'It' was too big, too complicated, too overwhelming, too distressing, too heartbreaking and too inspiring to put into words. We clammed up.

Years passed, and while we talked about having been in Darwin during Cyclone Tracy and working on the clean-up, few of us spoke in any detail about those experiences. Our memories were hard to relive, but I thought they were important. I thought there should be a record of our part in such a historic event in Australia's history.

In 2013, my dear friend Ruth Turnbull and I spent a weekend together before she passed away after a long and courageous battle with cancer. We set up a recorder on the table beside us and chatted for hours about our life in the WRANS while Ruth's wonderful husband John Clarke brought us endless cups of tea. We laughed about all the antics we had gotten up to. We cried over Cyclone Tracy and the effect that working on the clean-up had on us and our friends. We cried with mirth again, recalling our subsequent misadventures.

Ruth and I knew that everyone who'd been in Darwin at the time would have stories worth telling too. What was being done about that? I read a book on the history of the WRANS and only one paragraph mentioned that Wrans were in Darwin when Cyclone Tracy hit. A couple of books on the history of the RAN devoted only three paragraphs to the event and the Navy's response. Something needed to be done.

The Australian cattlewoman Sara Henderson once said, 'Don't wait for the light to appear at the end of the tunnel. Stride down there and light the bloody thing yourself!'

Righto. If I wanted a book about the WRANS and the RAN during Cyclone Tracy and the clean-up, then I'd have to write it myself. Writing a book couldn't be too hard. Lots of people did it. I'd have a red-hot go.

How could a woman of my years and experience have been so naïve? This has been yet another one of life's lessons hard learned.

My mother had saved all the letters I sent to her, and they were excellent memory joggers. I then researched official reports and documents, which provided the bare facts: dates, times, actions and statistics. I quickly realised that I would need to confine the book to the RAN's involvement, as what I knew about the Australian Army and the Royal Australian Air Force could be written on a postage stamp.

After that, I recruited the former Wrans and sailors I'd kept in touch with, so they could contribute their stories. That done, I put out the word for other former naval personnel, their dependants and civilian employees to provide first-hand accounts of Cyclone Tracy and the subsequent Operation Navy Help Darwin.

I was able to meet with some of the contributors face-to-face and record our interviews over cups of tea. Border closures and lockdowns due to COVID-19 made travel out of the question in most cases, and I conducted many telephone interviews in my little study, with my dogs snoring happily under the desk. Other people emailed their stories to me. Everybody gave their time generously, sharing with me their memories of that extraordinary period in their lives.

Few of the contributors had written about their experiences before. Few had talked in any detail about the cyclone and the clean-up apart from 'I was there' or 'I worked on the clean-up'.

The official documents, objective reports, have been available online from sources such as the Australian War Memorial and the National Archives of Australia for years.

Now, many of us are telling our own stories, providing our subjective human impressions. After all this time, many memories of the cyclone and the clean-up are hazy. That period of our lives passed in a blur, but some events made an indelible impact. It isn't important that we can't remember dates, times or places precisely; that's what the official reports are for. What's important is what we did, what we saw and what we felt. Our impressions and the vivid memories that have stayed with us bring the cold facts to life.

Readers won't be able to hear the wind sounding like a freight train passing directly overhead. They won't be able to feel the sting of the rain on their skin or feel the pain of muscles straining and bracing. They won't smell the rotting food and decaying bodies. After all this time, however, this is our attempt to answer the question, 'What was it like?'

PART 1

LIFE AS A WRAN

'Gentle ladies, you will remember till old age
What we did together in our brilliant youth!'

Sappho

CHAPTER 1

RECRUIT TRAINING, HMAS *CERBERUS*

Rain blasted horizontally across the room. In all my nineteen years, I had never seen rain flying sideways.

The cyclonic wind forced the torrential rain through the gaps in the louvres, flooding my room on the first floor of the Wrans quarters. I was soaked to the skin and the air was so thick with moisture that breathing was difficult. I took shelter in the built-in wardrobe, pushing my clothes, naval uniforms and shoes aside to make room.

Surely, this couldn't get any worse.

The fire alarms went off.

Sod all our training to respond instantly to alarms. There was no way there could be a fire with all this rain. Sod doing my duty to investigate. I was not going into that hell outside. No way. Not me.

That said, shuddering with apprehension, I climbed out of the wardrobe and waded through ankle-deep water to the door. The wind was fiercer out in the hallway, and water raged in rapids to the stairwell at the far end of the building.

I struggled against the wind along the corridor. I gripped the floor-to-ceiling metal louvres on the cabin walls and pulled myself along, desperately trying to stay on my feet.

Clearly there were no fires and evacuating the building was potentially suicidal. The alarms were most likely caused by an electrical fault.

The wind almost knocked me down when I crossed to the phone on the far wall, rang the duty Quartermaster Gunner and reported the alarms. We had to shout at each other to be heard over the roaring wind. He said to ignore the alarms. Thank goodness. I had never been so pleased to obey an order.

I crossed to the other side of the corridor and headed to my cabin with the wind at my back. A particularly strong gust knocked me off my feet and the water swept me along the corridor, feet first. I grabbed at the metal louvres along the walls, desperately trying to stop, and finally succeeded after sliding about ten metres. A ceiling fitting crashed at my feet. I lay on the floor gripping the louvres, being buffeted by the torrent of water flowing down the corridor as I tried desperately to control the panic. How much worse could this storm get?

* * *

Less than two years earlier, I had joined the WRANS, the Women's Royal Australian Naval Service. When I was in Grade 5, my sister Sandra joined the WRANS. She wore her uniform when she came home on leave and she looked so wonderful that I immediately decided to join up too when I grew up. And that's what I did, signing up on 23 February 1973, aged seventeen, straight after matriculating from high school. Like my sister, I was called up as a radio operator.

The previous year, 1972, had been momentous. We were thrilled to see the American Mark Spitz win seven gold medals in swimming at the Munich Olympics. But then we were horrified when Palestinian terrorists murdered an Israeli coach and athlete in the Olympic Village and took nine other Israeli athletes hostage, before killing them too.

Australian troops were still involved in the Vietnam War. The US President, Richard Nixon, was embroiled in the Watergate scandal. Ageing hippies were still singing 'Kumbaya'. Flared trousers and platform soles replaced kaftans. Gough Whitlam replaced William McMahon as Australia's Prime Minister.

Who knew what 1973 would bring? It started well for me as I travelled by train, an old rattler, from my hometown Rockhampton to Brisbane, where we Queensland Wran recruits were sworn in. We were all strangers from various towns in Queensland and were on our best behaviour. All this would change over the next few weeks as we all got to know one another, relaxed and let our personalities shine. After signing up, our group travelled by train to Sydney where we met up with the New South Wales recruits and had lunch at HMAS *Watson*, which must have been on the most picturesque site of all naval establishments. The back wall of the chapel had a huge glass window, giving worshippers a magnificent view of Sydney Harbour.

We Queensland and New South Wales recruits travelled by train from Sydney to Melbourne, and then there was another train to Frankston in Victoria. Then finally, we took a bus to the naval training base HMAS *Cerberus* on Westernport Bay, where we met up with recruits from all around Australia. We were a tired, motley bunch of young women. When we disembarked from the bus, we were lined up by a Wran reception party who tried to form us into a cohesive squad. A member of the reception party told us we were going to be assigned cabins. Cabins? Were we really going on board a ship on our first day? I didn't think women went to sea.

We discovered that the terminology used on the Navy's shore bases was the same as that used on board ships. Our bedrooms were referred to as cabins, floors were decks, walls were bulkheads, toilets were the heads and leaving the base was going ashore. Having cleared that up, we settled into our spartan accommodation in the Wrans quarters, Grant Block. This was a two-storey red brick building with no pretentions of style and elegance. The only decorative features were the anchors painted on the fire extinguishers.

I was assigned a small room, which contained two bunk beds against one wall and four small wardrobes along the opposite wall. We were each issued white sheets, a pillow and a blue and white bedspread with a picture of an anchor in the centre. Every morning, the beds had to be made up using hospital corners on the sheets and the bedspread needed to be positioned correctly. If our instructors found anything amiss when they did their daily rounds of the quarters, we would have to remake our beds to their satisfaction.

Two of my roommates were in my recruit Wrans intake. I had met Lyn Alderton, a fellow Queenslander, when we signed on in Brisbane. She was a blue-eyed blonde who quietly and intelligently took in everything around her. Joan Czarnecki, a bright young woman from Adelaide, South Australia, had short brown hair and soft brown eyes. She was of Polish descent and said that her schoolfriends could never pronounce her actual first name and called her Joan instead. I liked them both immediately. Our fourth roommate was a recruit Wran from the intake prior to ours. I couldn't have asked to be in better company. Lyn, Joan and I are still in touch after all these years.

An instructor told us that '10As are the rig of the day'. I surmised that she was speaking English but had no idea what she was talking about. She explained that our sets of uniforms were called rigs and were numbered for ease of identification. 10As consisted of a white short-sleeved shirt, navy blue skirt, beige stockings or white socks, black leather shoes and a white cap with a black tally band.

We were just getting our heads around all this new terminology when we discovered the slang. Soft drinks were called goffas, meals were scran, and showering was having a dhobey. On top of all our other training, we had to learn the whole new language of the Navy.

Then there was the all-important rank structure, which had to be studied carefully. There were not only Royal Australian Navy ranks, but their equivalent in the Australian Army and the Royal Australian Air Force. We had to salute officers and call them sir or ma'am, or we'd be in deep trouble. If we saluted petty officers and called them sir or ma'am though, we'd be bawled out loud and long and told never to do that again.

I spent the first week in such a state of bewilderment and confusion that I just wanted to sit down and cry, but there was never enough time for that. We were kept busy from 'wakey, wakey, wakey' at 0600 (6 am) to 'lights out, pipe down' at 2230 (10:30 pm) every day.

Eventually we learned enough to get by. Speaking the same language as everybody else in the Navy gave us a special feeling of belonging and we loved it.

* * *

During our training, we learned some of the history of our organisation. The WRANS was established in April 1941 when the RAN enrolled fourteen women at HMAS *Harman*, the wireless telegraphy station near Canberra. The WRANS was a non-combat branch of the Royal Australian Navy that alleviated manpower shortages resulting from men being assigned to combat roles during the Second World War. We learned that WRANS, upper case, was the Women's Royal Australian Naval Service and that the women who served in it were called Wrans, lower case.

We were told that the Wrans were there to 'release' the sailors so they could go to sea, not 'relieve' the sailors. The correct terminology was strongly stressed!

I really wanted to go to sea and was disappointed that women weren't allowed to serve on ships.

The WRANS was disbanded in 1948 but re-formed in 1951 to help the RAN cope with more manpower shortages. By 1959, the organisation was incorporated as a non-combatant (non-seagoing) part of the permanent naval forces. Women were finally permitted to serve aboard Australian naval ships in 1983 and the WRANS was incorporated into the RAN in 1984, but that was after my time.

* * *

Our divisional officer was 3rd Officer Maxine Kupke. She was blonde, beautiful and elegant, and I had the utmost admiration for her.

For some reason, WRANS officers didn't hold the same rank as RAN officers but had the rank structure of the Merchant Navy. WRANS 3rd Officers were equivalent to RAN sub lieutenants, 2nd Officers were equivalent to lieutenants, and so on. They were addressed as ma'am instead of sir. Behind their backs, all WRANS officers were known as 'Ma'am Dragon'.

The cold weather steadily intensified at *Cerberus* while we were undergoing our recruit training and basic radio operator training. After being raised in the subtropics, the cold wind in the usually hot months of February and March came as quite a shock to me.

The Navy only had summer and winter uniforms. No concessions were made for spring and autumn. It wasn't winter until the Navy said it was winter. We poor Queenslanders froze in our summer uniforms, only allowed to add cardigans until the Navy officially changed to winter uniforms. We were convinced that the wind at *Cerberus* came directly from the Antarctic.

* * *

We were a big group and it took time to get to know one another, but some people stood out from the start. Ruth Turnbull from Sydney, for instance, was the tallest woman in our intake. She was slim, olive-skinned and had long, shiny black hair. She reminded us of the American singer Cher, but while Cher was languid and drawling, Ruth was a ball of nervous energy. She was direct and didn't have an ounce of tact but was one of the most steadfast and loyal friends you could ever have.

We had been issued with navy-blue overalls and, of course, none of them fit well. Short recruits had to roll up the trouser legs to stop them dragging on the ground. In contrast, the trouser legs of tall recruits only reached partway down their shins and were said to be riding at 'half-mast', like flags on a flagpole. Ruth was so tall that there were no women's overalls long enough to fit her. She had to wear men's, but they were too wide across the shoulders and they swamped her.

We wore our overalls and anti-flash gear – white cotton gloves and hoods – for the exciting challenge of firefighting and NBCD training: Nuclear, Biological and Chemical Defence. Fires at sea are extremely serious and our firefighting training was comprehensive. At one stage, an instructor told our group that we were going to put out an oil fire with water. I didn't think that was possible, but the instructors gave us a two-gallon fire extinguisher each and told us the procedure to use them. Then two at a time, we had a go.

The nozzle of the hose of one extinguisher was set on 'spray', extending out over the oil fire and protecting the two of us from the flames and heat. The nozzle of the other hose was set on 'jet' and was trained steadily onto the far side of the fire. It took a while, but the flames were eventually smothered by the spray from both water extinguishers, which deprived the fire of oxygen.

From that, I learned that when the Navy said that something could be done, it could. When a lawful order was issued, it was entirely possible that the order could be carried out successfully. Less than two years later, the Navy told us that we were going to clean up Darwin after Cyclone Tracy devastated the city, and that's exactly what we did.

Wrans were not required to undergo all the training that the sailors did; not surprising, because at the time we were not allowed to go to sea. Occasionally though, we were given the option to perform some of the sailors' training. One option I chose was to navigate through a smoke-filled building. Recruits were required to wear a gas mask, enter through one external door, make their way through a maze of dimly lit, smoke-filled corridors and exit through another door at the other side of the building. I put on my gas mask and entered with some trepidation. The smoke was dense. I dropped low and slowly made my way along, but the smell was acrid. As it worsened, my air passages increasingly stung, so I gave up and returned to the entrance.

Once outside, I told the instructor my gas mask was leaking and letting the smoke in. He brusquely said that the seals on the mask were fine.

'But I could smell the smoke,' I wailed.

He roared, 'The rubber and leather of the mask smell because they've absorbed the smoke!' And he sent me packing.

I later discovered that I'd been three-quarters of the way through the building when I turned around and came back. By the time I exited, I'd covered one and a half times the distance required. Clearly, I was more than capable of completing the course. By turning back, I made the task much more difficult for myself and failed.

From that, I learned that if many people have successfully completed a task before me, I can do it too. And that if I commit to a task that's worth doing, it's usually much easier in the long run to see it through. That, too, kept me going through the worst of the clean-up after Tracy.

A couple of years ago, I mentioned the gas mask incident to Joan, and she said, 'The masks *did* leak. I was coughing up orange stuff for days!'

* * *

Our swimming test was a comedy of errors. We had to don our overalls, dive into the twenty-five-metre indoor pool, swim one and a half laps, tread water for five minutes or so, take off the overalls while still treading water and then swim to the end of the pool.

I had done lifesaving training at school, and we had always been instructed that if we were wearing heavy clothing we should do a star jump into the water and then swim breaststroke because freestyle is too tiring. When the whistle sounded, I did a spectacular star jump into the water and surfaced to see that the rest of the squad had dived in and were tearing down the pool doing freestyle! I was being left behind! Sod lifesaver training. I took off after them with a furious freestyle, belting down the pool. The sleeves of my sodden overalls rode up and down my arms with every stroke.

Ruth found her overalls a tremendous hindrance. She started swimming freestyle but said, '[I] was just going flop, flop, flop. In the end, I had to do breaststroke. It was the only way I could do it.'

Afterward, we went into the changing room to dry off and put on our uniforms. Few of us, including Ruth, had ever seen men's urinals before and had no idea what they were. Ruth entered, saw the urinal and thought, *Hey, this is really handy.* She began washing her feet in it.

Horrified, Lyn asked, 'Do you know what you're doing?'

'Washing my feet.'

'But you're in the men's urinal!'

'Oh, shit!'

After that, Ruth spent several anxious days waiting for tinea or some other exotic tropical disease to develop.

Our usually competent and capable Joan struggled in the water. She recalls:

> The swimming test! I couldn't swim and had to go several nights to try again and by the time you got back, the mess was closed, and you were lucky to get a bread roll. After that, I made sure I always had a packet of biscuits or something in the room just in case. I finally passed: doggy paddled the length of the pool and kept afloat.

* * *

Recruit Wrans had to stay on board for the first few weeks of training and spent Saturdays doing 'clean ship'. We wore our overalls and had to clean everything by hand. After finishing our assigned tasks, we gathered on the lawn and waited until our work was inspected thoroughly by WRANS officers and petty officers. If any fault was found, the offending recruits had to go back and clean their assigned area properly while the rest of us waited. We all just wanted to get it over and done with. Everyone was annoyed by offenders, so the standards of cleaning improved quickly.

Our class was particularly unfortunate because the Director of WRANS (also known as the Holy Chief) visited *Cerberus* on her last inspection tour before retiring. The cleanest base every year earned a trophy, and we were told that *Cerberus* had won it for decades. As this was the director's last visit, the pressure was on to make sure that we won the cup again.

Cleaning our cabins and the corridors in the Wrans quarters was particularly hard work. Firstly, we scraped back all the polish on the floor with knives. Then the floors were scrubbed and mopped, at least twice, then we had to lay polish down on them again. Most of this was done on our hands and knees.

We received inoculations for just about every disease known, even ones like smallpox, which had been eradicated. Some had nasty side effects and I think it was the one for yellow fever which had the worst effect on us. At least one of our recruits was hospitalised. We still had to clean ship for inspection, and we were dropping like flies from sickness and exhaustion while scraping back the polish on our corridor floor. Only one Wran recruit was left to finish the job of polishing our section and she earned our eternal admiration and gratitude.

We had been issued with stockings, which we never wore. We all preferred pantyhose, but the nylon stockings were useful for laying polish on the floors and shining our black leather shoes. We used to spit polish our black leather shoes and buff them afterward with stockings, which gave them an extra gleam. We were proud of our shoes and thought they were beautiful.

Our lectures were interesting and sometimes intense.

I wrote to my mother: *We had a security lecture the other day. Talk about James Bond stuff. We feel rather frightened now about talking about anything. And we don't even KNOW anything yet. Honestly.*

* * *

After a few weeks, we were given weekend leave and travelled to Melbourne by train. At seventeen, I was still underage. I needed written permission from my mother to go to the city and had to stay at the YWCA. That still seems bizarre to me. Even so, I had a marvellous time. Coming from a large but regional town, I was completely overwhelmed by the size of Melbourne. I even managed to get lost in the Myer department store. I went in one entrance, being careful to take note of my surroundings, but two hours later emerged from the store on a different street. Who knew the shop covered a whole block, back and front?

Ruth and a few other recruit Wrans were staying in the Southern Cross Hotel when there was a train strike, and they couldn't get back to *Cerberus*. For some reason, the *Cerberus* bus couldn't come to pick them up, so the Southern Cross put them up for three days, providing breakfast, lunch and dinner. As Ruth recalls, they 'had a ball'.

At one stage, they were in a shop that had a mirror outside and saw a Navy officer passing by. *Oh God, there's an officer*, Ruth recalls thinking. They decided to wait inside until he went past so they didn't have to salute but hadn't realised what they'd seen was his reflection in the mirror.

'We thought he'd gone past, but we walked out straight into him, so we saluted. We didn't know what else to do and he just laughed. I don't know who he was,' she says.

* * *

The parade training was hard, but in many ways it was the most memorable aspect of our recruit training. We were divided into two squads, one instructed by a petty officer (PO) and the other by a leading seaman (also known as a kellick). The chief instructor was Chief Petty Officer Quartermaster Gunner Berry. He was a tall, imposing man with a fierce red beard, and he had a booming voice that could paralyse a recruit at fifty paces.

One day, when we were marching along during training, Ruth's bra strap came loose and fell down her right arm. She was only able to swing her left arm and the PO charged up to her and bellowed, 'Recruit Wran Turnbull! Swing your arms!'

'I can't, sir, my bra strap has fallen down my arm.'

As they marched along, the PO tried to pull up Ruth's bra strap, all the time berating her for calling him 'sir'.

Ruth and I speculated this was probably the first time he ever tried to help a woman put her bra on instead of taking it off.

When we attended Divisions, which was the big parade involving all the troops on the base, our class didn't split into the two squads, but stayed in one big squad. The Wran officer called the orders but in the general din, Ruth, the right-hand marker and the front half of the squad didn't hear one of the commands. They marched straight ahead, while the rear half did a left turn, leaving the instructors scrambling to get us all together again.

We must have been a sad trial for our instructors.

* * *

The most lasting lesson of our recruit training was the importance of mateship and teamwork. At any given time, it seemed at least one of us was suffering from homesickness, tiredness, illness or felt overwhelmed. The friendship and support of our fellow recruits eased our anxieties, and we became a cohesive and supportive group. Our most enduring memories are of smiling faces and laughter. We Wrans developed a strong bond and that's what helped us cope, two years later, with the devastation of Darwin. We faced it together.

* * *

On 30 March 1973, our recruit training finished. Our service record states:

> This course is designed to train personnel in the basic principles of naval discipline, routines, conditions of service affecting a junior Wran, parade training, first aid and firefighting. It also encourages the development of

high standards of personal behaviour, hygiene and self-discipline.

The course includes:

General Service Training	43 hours
Parade Training	23 hours
First Aid	14 hours
Firefighting	7 hours
Communal Living	23 hours
Course Administration and Sport	43 hours

When we finished our basic training, we had a passing-out parade and I was awarded the prize for 'The Most Outstanding WRANS Recruit' of our intake. Without false modesty, this came as a shock to me. I had never topped my class or won so much as a book prize throughout my entire twelve years of schooling. I didn't top our recruit class in any subject either, even though I did my best, so being presented with a lovely pewter candy dish engraved with *Presented to Patricia Collins as the Most Outstanding WRANS Recruit – Class 140.1973* was unexpected and surreal. Perhaps it was my love for the WRANS that tipped the scales in my favour.

CHAPTER 2

COMMUNICATIONS TRAINING

After graduating from Recruit School, our intake was divided up for training in our assigned specialties such as sick bay attendants, stewards, writers and cooks. We recruit Wran radio operators transferred to the Communications School at *Cerberus* and undertook basic radio classes, which included Morse code. Some of us loved Morse, and the rhythm of it suited me well. Lyn and Ruth were to be in comms like me. I thought Joan would have done well as a radio operator but she was recruited to be a steward.

The Communications School was a complex of old wooden buildings on the far side of the base to our Wrans quarters. We marched there in the mornings, marched back to the quarters for lunch, marched back to comms school for the afternoon lessons and marched back to the quarters at the end of classes for the day. We had to line up smartly in the squad at the appointed time, stand waiting for further orders, then march smartly to our destination, where we again waited for further orders. This was typical of the armed services routine of 'hurry up and wait'.

It wouldn't have been so bad if we'd had time to eat our meals. Often, though, by the time we arrived at the mess for lunch, the cooks were starting to shut down the servery. It was against the rules to remove food from the mess, but we would quickly butter bread rolls, wrap them in serviettes and hide them in our hats or handbags to eat later, then race outside to form up our squad.

Despite the odds, we became reasonably good at marching. We worked out which was our right hand, which was our left hand, kept in time and avoided square gaiting. Mostly. Then the sport was to see how much we could get away with. When we marched to the comms school one day, Ruth, who started out as right-hand marker at the front of the squad, ended up down the back. We swapped all the way, with the one in front missing a step and the one directly behind doing a double step, all without being caught out.

Another day, as our squad was marching sedately down a narrow road at the comms school, a squad of recruit sailors came pounding around the corner toward us, doing double time. The guys in front were huge. We parted like the Red Sea, scattering right and left and flattening ourselves against the surrounding buildings, while they thundered past us. Our supervising Wran officer was incandescent with rage and gave us a furious dressing down for 'breaking ranks'. We took the criticism on the chin. Being trampled underfoot by a squad of sailors was not mentioned in our Conditions of Service.

The restrictions on us as recruits loosened a bit, and we had more freedom to go to the movies and to a few dances at the base. We had to wear our uniforms to the dances at *Cerberus* and, by this time, we were blessedly in our winter uniforms. The hall was always packed with recruit sailors and Wrans, and we danced nonstop to a great variety of music, old and new. The classic 'Rock Around the Clock' by Bill Haley and his Comets had everybody up and singing along. We always kept clothes brushes in our handbags as our uniforms had to pass scrupulous inspections at any time. Under the fluoro and strobe lights, the fluff showed up on our jackets and, as Ruth says, 'Everyone thought they had dandruff and they're going brush, brush, brush.'

We were also allowed to personalise our cabins with our own sheets and bedspreads, although our beds had to be made with the navy-issue sheets and bedspreads for weekly inspections.

At the end of our basic radio operator course, we were again divided up for further training as radio operators teletype and radio operators Morse. To my great delight, Lyn Alderton, Ruth Turnbull and I were chosen for the latter category.

* * *

The ten of us selected for further training in Morse were posted to HMAS *Harman* in Canberra. *Harman* was affectionately known as the 'Stone Frigate' because it was far inland, not on the coast where you would normally expect to find a naval base.

We spent the entire winter there. The brittle dry cold in the ACT was different from the harsh Victorian winds, but just as hard to endure.

Our rooms in the Wrans quarters at HMAS *Harman* were divided in two by a partition wall, with two beds on one side and two more on the other. We were issued the inevitable white sheets and blue-and-white Navy bedspread but were allowed to use our own bed linen and could even put posters on our walls. One of the young women in my room put up a romantic black-and-white poster of Olivia Hussey and Leonard Whiting in the 1968 movie *Romeo and Juliet* and we loved it.

Heating was supposed to be provided by column heaters connected by water pipes, but they rarely worked. I'd go to bed every night wearing a spencer, flannelette pyjamas, dressing gown and bed socks and wrap myself up in a couple of blankets. Yet I would still wake up shivering during the night.

In the depths of the Canberra winter, we Queenslanders would rug up in as many clothes as we could possibly wear.

For work, this usually constituted bra and panties, singlet, spencer, two pairs of pantyhose, petticoat, white uniform blouse, tie, serge skirt, navy blue woollen vest, navy blue cardigan, black serge jacket, navy blue Burberry (overcoat), black leather shoes and brown leather gloves, all topped with a white cap. With our little spindly legs topped by so many clothes, we looked like walking mushrooms.

From Monday to Friday each week, we'd travel by Navy bus from HMAS *Harman* to the Navy radio station and training facility in the middle of the Bonshaw sheep property. Bonshaw Receiving Station was an old wooden building with no insulation and inadequate heating. Our leather gloves were thin and provided little warmth, so our hands were stiff with cold by the time we entered our training room. That made it hard to write or type, so we'd take turns standing on chairs and thawing out our hands in front of the strip heaters mounted high on the walls. In the communications branch, petty officers were known as radio supervisors (RS) and our instructor, RS Lyn Hinge, would give us a stern lecture if she caught us, citing the probable development of chilblains. Lyn was short, slim and dynamic. Whenever she entered our classroom, she would quiet our chatter by telling us to 'keep it down to a dull roar'. I thought we were very lucky to be trained by her.

A civilian, Jerry Albert, taught us the technical aspects of radios. Unfortunately, the science went right over my head. I loved operating radios but had no idea how they worked, no matter how carefully Jerry explained the phenomenon.

When Ruth became extremely tired, Jerry was the one who realised she was unwell and urged her to be checked out by the Navy doctor. She was diagnosed with glandular fever and given the option of being admitted to hospital on the RAAF Base or going home to Sydney for six weeks to recover. She said, 'So what self-respecting person's going to go to the hospital in the RAAF Base?'

The inter-service rivalries were strong!

So, she went home. That put her a long way behind in the Morse training, and she felt that she never quite caught up. Alongside the rest of us, she still passed all the exams, however, and we were very relieved and pleased for her.

Sport was played every Wednesday afternoon. There were tennis courts on base and some of us really tried to play well. My balls often lobbed over the high fence though, and I swear I spent more time walking around picking up my balls than hitting them back over the net. Another activity was a cross-country run out to the Bonshaw Receiving Station and back. A couple of us would head out from *Harman*, trotting, pretending we were just warming up, and would drop to a walk as soon as we were out of sight. We strolled along, trying not to startle the sheep and chatted happily until we were back in sight of the base, when we would work up to a trot again. I doubt anybody was fooled. My childbearing hips and lumpy legs were not made for running. I loved swimming though and if the base had a swimming pool, I would have trained there every day. The best I could do with the other sports available was to turn up and, as the Brits say, 'show willing'.

The radio operator Morse course was the longest of all, so, although our training would continue until November, we picked up our rate in August, the same as the rest of our recruit Wran intake who had finished their training and were considered qualified as stewards, cooks, sick bay attendants and so on. We were now classed as radio operators and our new rank was Wran – the equivalent to an army private.

We eventually finished our training on 8 November 1973 and then the category was changed to Radio Operator Special. Of course, we preened over our new designation. The Powers That Be insisted that the word 'special' applied to the work we did, not to us personally, but we happily ignored them.

The year 1973 had been huge for us and was also a cultural touchstone. The movie *American Graffiti* starring Ron Howard,

Richard Dreyfuss, Cindy Williams and Harrison Ford was released, as was *The Sting* starring Paul Newman and Robert Redford, which I particularly enjoyed. The Sydney Opera House opened, and Princess Anne married Captain Mark Phillips. On the political front in 1973, the last Australian and US troops were withdrawn from Vietnam, but there was much unrest still around the world. The Irish Republican Army (IRA) bombed King's Cross and Euston train stations, Whitehall and the Old Bailey in London, and Manchester City Centre, but a referendum in Northern Ireland still resulted in a vote to stay part of the United Kingdom. The Yom Kippur War started when Egyptian and Syrian troops attacked Israel.

I don't think there was a television in the Wrans quarters at *Cerberus*, but there may have been one in the recreation room or the Junior Rates Club at *Harman*. In any case, I can't remember watching any TV shows. Some of us listened to little transistor radios to keep up with new music and current events. My downtime was spent playing sport, going to the movies, window-shopping, reading books and sightseeing. The Australian War Memorial was fascinating, and I went there a few times. I had learned a bit about the Great War and the Second World War at school but hadn't seen the uniforms, weapons, military vehicles and aircraft before now. To me, seeing equipment that had actually been used in those conflicts made the dry textbooks come alive. I was proud to be serving my country, even though I had no expectation of experiencing war.

When our training finished on 8 November 1973, we were fully qualified and ready to take on the world. I was to have a couple of weeks leave in my hometown, Rockhampton, and then travel to Darwin to undertake watchkeeping duties. I could hardly wait to get going and packed my belongings in record time.

CHAPTER 3

DARWIN

After a couple of weeks leave in November 1973, I travelled by plane to HMAS *Coonawarra* in Darwin in the Northern Territory. The heat and humidity hit me the moment I emerged from the aeroplane at the Darwin Airport. The tarmac rippled in a haze of heat waves. It was worse than Rockhampton had ever been. I'd joined the WRANS eager for new experiences but hadn't reckoned on going from a frostily cold Canberra winter to the heat and humidity of Darwin in the wet season.

Ruth Turnbull and another Wran were in Canberra Airport, in uniform, preparing to depart for Darwin when they saw a group of officers walking toward them. Being conscientious young Wrans, they stood to attention and saluted, only to realise the officers were pilots in the commercial airline TAA. The delighted pilots walked past them, grinning from ear to ear.

I was assigned a cabin in the so-called 'new' Wrans quarters at HMAS *Coonawarra*. This was a three-storey brick building with accommodation on the first and second floors, and a recreation room, laundry and undercover car parking on the ground floor. Joan Czarnecki remembers the goannas we would encounter on the stairs of the Wrans quarters: 'Some were huge; you ended up using the other stairs until they left.'

The old Wrans quarters was a two-storey building across the lawn from the new one, with the sick bay located on part of the ground floor. We had a bed, wardrobe and desk each. A ceiling fan was positioned over each bed and the foam mattresses were covered in plastic – very necessary considering how much we would sweat in the heat and humidity.

After showering, we would dress under the whirling fans, otherwise our clothes would stick to our sweaty skin and be almost impossible to pull on. We were permitted to wear brown sandals with our working uniforms, instead of shoes and socks or pantyhose, thank goodness.

A great joy for me was having a twenty-five-metre swimming pool on the base, just over the paddock from the Wrans quarters. I loved going for a swim to keep fit and cool off after work or on days off.

The beaches around Darwin were long and beautiful, but swimming in the sea was prohibited from October to May every year due to the dangers from box jellyfish. Even paddling in the shallows was risky and many adults and children were stung in ankle-deep water. Adding to that peril were the saltwater crocodiles, huge monsters whose territories encompassed not only the sea but also the great fishing spots inland.

* * *

The headquarters of the Royal Australian Navy North Australia Area was in Darwin and Captain Eric Johnston was the Naval Officer Commanding (NOCNA).

The two shore establishments were located on unlikely sites. HMAS *Melville*, also under the command of Captain Eric Johnston, occupied quarters at the Army's Larrakeyah Barracks. *Melville* supported the patrol boat base and provided some logistical support for visiting ships.

HMAS *Coonawarra* started life as a wireless telegraph station, which was built several kilometres inland from Darwin during the Second World War. When the war ended, the station continued its communications role. In 1954, the telegraph station was renamed the Darwin Naval Radio Station. It wasn't until 16 March 1970 that the area was formally commissioned into the Royal Australian Navy as HMAS *Coonawarra*, under the command of Commander Williams. By that time, according to Commander Williams, the wireless telegraph station was 'held up only by virtue of the bodies of countless dead insects'.

Fortunately, new communications facilities were under construction at Shoal Bay. In the meantime, sailors, Wrans and Army personnel from 71 Signal Squadron worked together at the old radio station, along with civilians such as Michael Pilkington. Michael and his wife, Christine, moved to Darwin in 1971 when he was appointed to the position of technical officer working with the Department of Supply, Navy.

Michael, along with the Navy techs, maintained all the receivers at *Coonawarra*'s radio station. We Wrans were usually at the top of our game when operating the radio sets but knew the bare minimum about their repair and maintenance. That's where the technicians came into their own.

Michael says, 'I soon got to realise that what the Wrans were saying about the radios was usually right. If something was wrong, they'd tell you. They were working on the damn things eight to ten hours a day, so they must know what was going on.'

Michael and I both worked at the old radio station. I met up with him and Christine in 2020 and we reminisced about our time in Darwin. Christine says that when Michael was offered the job in Darwin, he 'actually got two jumps up in grade, so he couldn't say no really'. Christine had just found out that she was pregnant and the move to Darwin, although good for Michael's career, was challenging. They later lost their son when he was only two days old, and he was buried in Darwin.

* * *

When our little group of newly qualified radio operators arrived in November 1973, HMAS *Coonawarra* was under the command of Commander Stanley A Hall, and it operated from two sites on the Stuart Highway, *Coonawarra* West and *Coonawarra* East.

We were very excited to be qualified radio operators at last and relished being on watchkeeping duties. At first, our shifts consisted of two 1600 to midnight shifts, called the first watches, followed by two day watches from 0800 to 1600, followed by two days off. We also took turns at a 0900 to 1700 shift, Monday to Friday for a week at a time.

I was amazed at how much free time we had. It was a bit hard to get together with so many people on different shifts, but we managed lots of 'smokies' (parties), especially after the initial first watch, which finished at midnight. There was certainly no time after the second first watch to party, as we had only eight hours to travel, bathe, sleep and have breakfast before the first day watch.

Despite the grief of losing their newborn, Michael and Christine Pilkington also loved life in Darwin and made many friends. Christine says, 'I tell you what, they were a damned good mob. We could have partied every night. We tried in the beginning, but we couldn't keep up. We had two young boys then, two sons you know. We tried, but we couldn't do it.'

We Wrans were young, fit and single and had no trouble keeping up! Our free time was generally filled with activity. Most sailors and Wrans played sport, often in local civilian competitions. I swam most days and played squash regularly with my Navy and Army mates. Mind you, I wasn't much good. Not being well coordinated, my squash ball ended up in the neighbouring courts a bit too often. Still, I enjoyed the game and the exercise.

The Stuart Highway was the only road out of Darwin and stretched to Adelaide in South Australia. Anyone travelling south from Darwin along the highway was said to be going 'down the track'. We loved spending the day down the track at Berry Springs or Howard Springs, splashing around and swimming in the cool water, fishing and picnicking. We had to be careful of the wildlife. Water buffalo and huge goannas abounded. Saltwater crocodiles thrived in the conditions and preyed on almost anything edible, including people. The crocs were between three to four metres long. I was constantly on the lookout for them.

We often went camping further down the track on our days off. The physical training instructors (PTIs) kept camping gear in their stores and would loan them out to us. There was no way we could sleep in tents in those hot and humid conditions, so at night we'd lay on top of sleeping bags on the ground with hoochies (plastic shade cloths) pegged over us. There was no such thing as 'glamping' in those days. By day, we'd cook our freshly caught barramundi on open fires and drink tins of beer. It was a great life.

The only fly in my ointment was the complete lack of fresh milk in Darwin. We had to make do with powdered milk, which most of us disliked. The cooks provided us with flavoured syrups such as chocolate and strawberry to mix with the powdered milk to make it more palatable.

There was a public telephone on the ground floor of the Wrans quarters, and I rang my mother once a week. We would all save our coins to put in the slot of the telephone every three minutes when prompted by the telephonist at the exchange. Sometimes, the telephonist might forget or be too busy to interrupt us and we'd talk for free for a few extra precious minutes.

Other than that, we kept in touch with family and friends by letter, although mail to Darwin wasn't always reliable. I once received a Christmas card from an aunt in March.

* * *

The Third Australian Patrol Boat Squadron (AUSPABRON3) was also based in Darwin. It comprised the Attack class ships HMAS *Attack* (pennant number P90), *Assail* (P89) and *Advance* (P83). When in harbour, the patrol boats normally berthed at either Stokes Hill or Fort Hill wharves, but they spent most of their time at sea.

Pennant numbers are used by the Navy to identify ships. Every ship in the Royal Australian Navy has a unique identifier that is painted in white on both sides of their hulls. Ships such as Attack class patrol boats look alike and have the same silhouette on the horizon. The number on the hull enables individual ships to be identified even from a great distance. In recruit training, we learned the pennant numbers of all the RAN ships and could pick them out at a glance.

My dear friend Dot Wright, who had been in my comms class, had friends on board HMAS *Attack* and I drove her to the wharf one day to catch up with them. I was very excited to be on a patrol boat, even as a visitor. I envied the sailors who were able to go to sea in them. While we were having a beer in the petty officers mess, a young PO emerged from the showers. He had not expected women to be in the mess and was embarrassed to be caught out wearing only a towel. He sped past us and exited the room, returning a short time later, fully dressed in his working uniform. I was smitten.

Bernie Kloss was exactly what he seemed: a country lad, fair-haired, young, capable, and enjoying the adventure of a lifetime. He was a leading hand, second engineer on HMAS *Attack*, which he joined at the end of July 1974. Bernie's younger brother's birthday was on 3 August, and Bernie phoned him then, just before *Attack* had to do a rushed sail down the West Australian coast to investigate illegal fishing activities.

The patrol boats were actively involved in controlling incursions by foreign fishing boats, mainly Indonesian, in

Australian waters. The foreign sailors dived for trochus shells, which they took back to Indonesia to make items such as buttons and watch faces. The little wooden fishing boats rarely had refrigeration and the sailors mainly lived on salted fish and dried foods, but some carried live chickens. The Australian government was not only concerned about the large-scale illegal fishing in Australian waters and drug running, but also about the threat of foot and mouth disease. The three patrol boats spent a lot of time at sea, patrolling from Thursday Island around to Port Hedland. They were effectively Australia's Coast Guard for almost a quarter of Australia's coastline. Typically, HMAS *Attack* would sail from Darwin and patrol the northern waters to Broome in Western Australia where they would fuel up and sail back to Darwin, usually passing another patrol boat en route.

The intensification of illegal activity stretched the capabilities of the three patrol boats and a decision was made to augment the Third Australian Patrol Boat Squadron with a fourth vessel, HMAS *Arrow* (P88).

Arrow had been a RAN Reserve patrol boat in Melbourne until early 1974. The reserves struggled to maintain and man the vessel and she was in a poor condition when the 28-year-old Lieutenant Robert 'Bob' Dagworthy, RAN, was appointed her new commander and a new crew was posted on board.

Coincidentally, this was not the first time Dagworthy had commanded *Arrow*. Before his patrol boat commands, he had served in the Daring class destroyer HMAS *Vendetta* (08), which finished a refit in Melbourne and underwent trials in 1973. Targets were required to be towed for gunnery trials and, even though *Arrow* was a reserve vessel, she was loaned to the permanent Navy for that task. As one of the watchkeeping officers on *Vendetta*, Dagworthy was given temporary command of *Arrow* and drove her out of the training base HMAS *Cerberus*. His executive officer (XO) for that task was Sub Lieutenant Russ Shalders, who later became Chief of the RAN.

After *Arrow* was taken over by the RAN, much had to be achieved in a short space of time. General repairs and maintenance were carried out, stores brought on board and properly stowed, and training and sea trials undertaken. According to Dagworthy:

> The whole ship's company came together superbly. They were an outstanding team. The command team included the energetic, positive executive officer (XO) John Jacobi. The Coxswain Barry Spencer was a 'rock', a superb leader, full of sensible ideas. The Charge Engineer Danny Grosse was an outstanding technician who was a perfectionist in managing the ship's engineering systems. The seamen on board were led by the ever reliable, cheerful and colourful Buffer, Les Catton, who sadly lost his life during Cyclone Tracy. Of all the ships that I have served on, I regard *Arrow* as the happiest, the most competent and the most unified team that I have worked with. I also remember Ian Rennie who we called 'Jack'. He also lost his life during Cyclone Tracy. Jack was one of the stoker team, a really solid citizen. He was quiet, but whenever work had to be done after secure [the end of the working day], Jack was first to put up his hand and was prepared to soldier on to get things completed. I remember whenever I kept a watch with him, he made a great egg and bacon sanga.

A complete work-up of *Arrow* was performed in Sydney, followed by a full inspection and an operational readiness evaluation before she sailed to Darwin in August 1974. *Arrow* arrived in Darwin on 2 September 1974, and Lieutenant Dagworthy reported to Captain Eric Johnston, who was the Naval Officer Commanding North Australia Area (NOCNA) in Darwin at the time. Johnston was known as 'The Big E' and was an excellent, charismatic and very professional leader.

Dagworthy and Johnston had served together on *Vendetta* for nine months when Dagworthy was a midshipman and Johnston was a lieutenant commander and XO. Johnston's nickname for Bob was 'Daggers'.

At this formal first meeting in Darwin, the discussion between 'The Big E' and 'Daggers' ranged to communication procedures at sea. When the patrol boats were out of UHF/VHF range, communications were limited to Morse code and had to be kept brief. Johnston stressed that Dagworthy was never to send a 'propose' signal. He directed Dagworthy to always send an 'intend' signal. A 'propose' signal is sent to request approval of an operation. An 'intend' signal means an outlined action will be commenced and continued unless countermanded. Captain Johnston commented that he knew Dagworthy and was confident Dagworthy would make the correct decision, and he would fully support him. From then on, Dagworthy only sent 'intend' signals and his decisions were never overruled.

In those days before computers and mobile phones, few naval families in Darwin even had home phones. Because Bob was in command of *Arrow*, he and his wife, Christine, had a telephone at their house, connected to HMAS *Coonawarra*. Sometimes when *Arrow* was on patrol, Bob would send a Morse signal to the Naval Communications Station, saying that at a certain time they'd be close to one of the commercial radio stations at Broome or Derby or Thursday Island and would be back in UHF/VHF range. He would arrange for the crew to have a radio telephone conversation with their wives when the ship was in range.

Christine Dagworthy would receive a message notifying her when they would be able to phone from sea. None of the other wives of *Arrow*'s crew had phones, so Christine would drive to their houses and let them know when they could come around to her house for the phone call. At the appointed hour, the wives would take turns speaking to their husbands on her phone.

'I have to say,' remarks Christine, 'those telephone calls were very, very welcome for us, as well as for the men on board. It was delightful.'

Christine was an invaluable conduit for communications between the ship's complement of *Arrow* and their families. She has many fond memories of those early months in Darwin. Christine was twenty-six years old when she flew to Darwin from Sydney in September 1974 with their daughters, two-and-a-half-year-old Siobhan and four-month-old Kate. Their dog, Germs, named after Germaine Greer, was flown up to Darwin on a commercial aircraft a few weeks later. Germs had been staying with Christine's parents and, after the Dagworthys were settled in the house, Christine's father drove Germs to Mascot Airport and saw her on to the plane. Christine then picked her up at Darwin Airport. She recalls, 'Germs was wagging her tail as she was offloaded from the aircraft in a wire cage.' In those days you could walk out on to the tarmac to welcome passengers and pets.

On occasions when Bob and Christine were invited out, John Jacobi, Bob's second in charge, would babysit the children. Christine came to know the crew of *Arrow* and found them a very friendly group. They all had a lively sense of humour, and she was very fond of them.

The *Arrow* ship's company considered it a great adventure to be on a patrol boat in the north, particularly when they covered the Kimberley Coast. In those days it was a wilderness and an area of stunning beauty.

* * *

All four patrol boats in the Third Squadron took part in SQUADEX 74. The exercise commenced at 9 am on

Wednesday 18 December when *Advance* and *Assail* cast off from the dock and proceeded out of the harbour in company. They were joined by HMAS *Attack* and carried out officer of the watch (OOW) manoeuvres, heaving line transfers, gunnery exercises and various seamanship evolutions.

HMAS *Arrow* joined at the completion of the gunnery exercises and the SQUADEX program continued as planned. The Australian Broadcasting Commission (ABC) was invited by NOCNA to come on board *Arrow* for the day to cover the exercise. During the night encounter exercise, *Arrow* took on the role of an enemy missile-firing boat and was successfully prevented from reaching Darwin.

The patrol boats weren't state of the art by that stage. For instance, the gun on the forecastle of an Attack class patrol boat was a Second World War anti-aircraft gun that had been based in London during the Blitz. They were totally inappropriate for patrol boats. Every time the gun was fired, the vibrations would burst the fluoro tubes in the forward part of *Arrow*.

Whenever Dagworthy advised the ship's electrician, Kevin Rainbow, that he was going to fire the gun, Rainbow would groan inwardly with exasperation. He recalls thinking, *Bloody hell, I've got to go and replace every fluoro in the ship after he does that silly business.*

Seamanship exercises and weapons training continued on Thursday 19 December. SQUADEX 74 was the first time that all four patrol boats of AUSPABRON 3 worked together as a unit, and was also, fatefully, the last.

In his report of proceedings for December 1974, Lieutenant CJ Cleveland, RAN, commanding officer of HMAS *Assail*, wrote that the 'Two days of evolutions, firings and drills ... proved to be of great benefit'.

The patrol boats anchored overnight and then berthed at Stokes Hill Wharf on Friday morning to embark personnel for a family day. Basic manoeuvres were carried out during the day. The ships returned to the wharf at 2:30 pm Friday afternoon to prepare for a quiet Christmas.

Michael Daly was a twenty-year-old Navy cook who had volunteered for service on patrol boats. When a vacancy for a cook came up in Darwin, he considered himself lucky to get the nod. He joined HMAS *Advance* eight days before Cyclone Tracy hit. He had barely settled in before the annual squadron exercise took place. He says:

> When we came back, we had a sort of a wash-up at the Don Hotel, and I met most of the other guys that night. I met Les Catton, *Arrow*'s buffer, and young Jack Rennie who was a stoker on *Arrow*. I was talking to Jack as we were at *Leeuwin* together, but he was the intake behind me. He was saying that he didn't like Darwin at all, and he had a posting south after Christmas, and he was really rapt to be going but, unfortunately, he never made it.

Michael was concerned about the cyclone season in Darwin, but his shipmates reassured him, saying, 'This happens every year. If a warning comes along, the boats will go out and tie up at the buoys out in the harbour.'

That Friday night, after the SQUADEX exercise, Captain Johnston listened to the sports program on TV as usual, mainly for the form guide and scratchings for the upcoming Saturday race meeting. He was a keen race-goer and later, when he was the Administrator of the Northern Territory, he owned racehorses. The Darwin sports editor had been in the ABC team that had covered SQUADEX 74, and he continually talked during the sports program about what a great day they'd had aboard *Arrow*. It was Captain Dagworthy this and Captain Dagworthy that! Dagworthy was indeed the captain of *Arrow*, but he was only a lieutenant, not a captain in rank. At the end of the sports program, Captain Johnston rang Bob Dagworthy at home. Christine answered the phone, and he said to her abruptly, 'I want to speak to Daggers.'

She said to Bob, 'I think you're in a spot of trouble. Captain Johnston wants to speak to you, and he sounds pretty gruff.'

Bob picked up the phone, 'Sir, Dagworthy speaking.'

Johnston sternly asked if Bob wanted his captain's shoulder boards and brass cap dropped around immediately or would he come round and pick them up from the Cell Block Bar the next day after the races. Dagworthy apologised. He said he should have explained to the ABC team that he was only a lieutenant, a commanding officer, not a captain. Johnston roared with laughter and then said that Dagworthy had usurped him, but that it was excellent PR – and hung up!

We radio operators were enjoying our work in Darwin too. The old wooden radio stations were run-down and the equipment dated, but a new state-of-the-art communications station was being constructed at Shoal Bay. According to civilian technician Michael Pilkington, the building of the new station was almost finished at the end of 1974 and the fitting out of the interior had begun. The floor tiles were being laid and the extensive wiring was underway. However, there were still many more months of work to be done before the building was expected to be operational.

* * *

The year 1974 had been eventful so far, not only for the Navy in Darwin but all around the world. In June, a bomb, suspected to have been planted by the IRA, had exploded in Westminster Hall, the oldest part of the British Houses of Parliament. In August came the resignation of the American president, Richard Nixon, and Vice President Gerald Ford was sworn in as his replacement. In October, The Rumble in the Jungle took place in Zaire, where Muhammad Ali knocked out George Foreman in eight rounds to regain the World Heavyweight title.

By December 1974, *Melville* and *Coonawarra* were manned by 305 naval personnel, including ninety Wrans and twenty-nine civilians. The combined ships' complements of patrol boats totalled seventy-five personnel. There were also 287 naval dependants resident in Darwin, so the Royal Australian Navy had a presence of 696 in a city whose population was approximately 47,000.

I loved everything about my new life: the work, the Navy, the WRANS, the sport, the recreation. I was quiet back in those days and, rather than being an instigator of recreational activities, I was happy to go with the flow and quietly absorb every minute of this great adventure.

On Christmas Eve 1974, Cyclone Tracy hit Darwin, and everything changed forever.

PART 2

CYCLONE TRACY

'The important thing is to accept that
some days are just for surviving,
and not to give up.'

Anon

CHAPTER 4

THE BUILD-UP TO TRACY

Between 1964 and 1975, twenty-five cyclones had been named in Northern Territory waters. Eleven approached within 200 kilometres of Darwin and resulted in gale-force winds and heavy rain, with minimal damage to infrastructure.

Those of us who were raised in the northern parts of Australia were familiar with the threat of cyclones during summer. We knew that tropical cyclones are intense circular storms that originate over warm tropical oceans and are usually around 320 kilometres in diameter. The winds swirl in a clockwise direction around a central region of low atmospheric pressure called the eye.

I and many other Queenslanders and Northern Territorians had often experienced the effects of the outer rings of cyclones, which generally have an outer radius of about 160 kilometres and an inner radius of about 30 to 50 kilometres. Damage was usually limited to fallen tree branches and flooding.

The second and most dangerous and destructive part of a tropical cyclone, however, is the eyewall. This region, about fifteen to thirty kilometres wide, is where winds are strongest, and rainfall is heaviest. The eyewall exhibits ferociously strong horizontal winds as well as updrafts, vertical windspeeds up to five to ten metres per second.

The winds are at their very strongest at the innermost edge of the eyewall surrounding the central region, called the eye. Within the eye itself, the air is often calm and the skies are clear.

In his dissertation, 'Operation Navy Help – Disaster Operations by the Royal Australian Navy post-Cyclone Tracy', Eric Johnston states that, 'It would not be unfair to say that despite many warnings the population of Darwin had a laissez-faire attitude regarding the likelihood of them suffering as individuals from such a natural disaster.'

There is always the attitude of 'it can't happen to me', but to be fair, few would have had any idea of the devastation wreaked by the full force of a cyclone.

On 3 December 1974, Cyclone Selma, which had formed in the Timor Sea, was only fifty-four kilometres west of Darwin. There were winds of ninety kilometres per hour (km/h) on Bathurst Island, but Darwin was only affected by strong winds and heavy rain. At noon, Selma stopped, reversed to the north-west, and then moved south-west away from Darwin, having only caused minor damage to the city. Several trees were uprooted. In the harbour, trawlers, which had been poorly secured, broke free from their moorings. Three were driven aground.

In my own happy ignorance, I found the persistent heavy rain of a cyclone more of a nuisance than a danger. The humidity, always high in the wet season, went up a notch and we always felt wet. The clothes and shoes in our wardrobes would develop persistent mould. I cleared out my wardrobe once a week, cleaned all the surfaces and then wiped them down with bleach. Our shoes were especially hard to clean of mould and we sprayed them regularly with Glen 20.

On 20 December 1974, the Bureau of Meteorology monitored the formation of a weak tropical low in the Arafura Sea, 700 kilometres north-east of Darwin. By the following morning, satellite photographs showed the large cloud mass about 640 kilometres north of Darwin. Continued monitoring by satellite and radar showed the mass making slow and steady progress to the south-west. The first tropical cyclone warning was issued at 4 pm Central Standard Time (CST) on 21 December. The depression proceeded to develop the typical clockwise cloud system of a tropical cyclone and, with winds passing sixty-three km/h, the cyclone was officially classified and named Tracy at 10 pm that evening.

For the next few days, Tracy was watched closely while it moved south-west. If the cyclone stayed on that course, it would not pose a major threat to Darwin as it would pass the city well to the north. We carried on with our Christmas preparations. I had done all my Christmas shopping and sent presents home a week earlier. All my cards had been written and sent and now I was counting down for the big night.

The tradition at *Coonawarra* was to have a dance at the Junior Rates Club on Christmas Eve, with a fancy-dress theme. There were no costume shops in Darwin in those days, so we scoured the charity shops and cobbled together imaginative, witty outfits. At the end of the dance, Wrans and junior sailors would sing carols around what was affectionately known as the 'marriage patch', houses on the outskirts of the base where naval families lived. A mob of young people, happily fuzzy with alcohol, singing loudly off-key while wandering down the road in the middle of the night would not have been welcomed by parents with children, but it was fun for us. I could hardly wait to do it again this year.

Another tradition was for the officers to serve Christmas lunch to the lower ranks. Being served roast meat, vegetables and plum pudding by naval lieutenants was an undreamed-of event, a real treat. Most officers performed the duty with good humour, but others showed poor grace. Still, they did it, and I thought it was a lovely tradition.

We had heaps to look forward to. Not only were we preparing for Christmas, but there was a lot of interest in sport.

Cricket was uppermost. The Australian cricket side, captained by Ian Chappell and including players such as Rod Marsh, Greg Chappell, Dennis Lillie, Max Walker and Jeff Thomson, were tough, aggressive competitors. The team was labelled the Ugly Australians for their antagonistic attitude toward the opposition, their foul-mouthed sledging and hostile bowling. Their skills and strength were impressive and Aussie cricket lovers revelled in the successes of the team. They beat Pakistan 3–0 and the West Indies 2–0 in the 1972–73 cricket season and also beat New Zealand 2–0 in 1973–74. For the 1974–75 season, they were playing against the touring England team.

Australia won the first Test in Brisbane and the second in Perth. The third Test was to be played in Melbourne starting, for the first time, on Boxing Day. Even people who didn't like cricket couldn't avoid the hype. I enjoyed watching Ian Redpath and Doug Walters bat and was looking forward to the TV and radio coverage of the game.

During an evening broadcast on 22 December, ABC news radio reported that 'Cyclone Tracy poses no immediate threat to Darwin'. Excellent. It would be business as usual for us at Christmas.

At 3:30 am on 24 December, the first direct evidence of the wind strength was obtained from the automatic weather station at ADV Cape Fourcroy, which recorded a mean wind speed of 100 km/h. At 9 am on 24 December, Tracy passed the western tip of Bathurst Island, north of Darwin, and changed direction, first southerly and then east-south-east toward Darwin. By 9:30 am, the eye was fifteen kilometres from the weather station at Cape Fourcroy, and the mean wind speed had increased to 120 km/h.

From then on, the cyclone intensified and travelled directly toward Darwin. That was going to put a dampener on our celebrations, but we were resilient and determined to carry on.

Captain Eric Johnston reported that on 24 December, 'I received a signal request from Marine Operations,

Canberra, to send a patrol boat to the assistance of a Japanese trawler reported to be in distress to the west of Bathurst Island.' Johnston was concerned about the lack of information regarding the fishing boat's precise location and the strong probability that the boat was in the path of Cyclone Tracy. He informed Marine Operations that he 'would bring a patrol boat to immediate readiness but would not sail it until a clearer picture of the weather in the area had been obtained'.

The Weather Bureau in Darwin hired a light aircraft to search the general area where the fishing trawler was thought to be located, but the pilot had to turn back because the weather was so bad.

At 12:30 pm (CST) on 24 December 1974, a top priority Flash Cyclone Warning was issued advising of Tracy's expected landfall early on Christmas morning. This was broadcast by ABC radio and the commercial radio station in the early afternoon and then at thirty-minute intervals. The broadcast messages contained not only warnings about the cyclone, its intensity, position and projected path, but also advice as follows:

> A cyclone is imminent. A few hours of organised family action can turn the odds your way. We present this advice from the Darwin Weather Bureau and the Northern Territory Emergency Services Organisation as a community service for Top End listeners.
>
> Make sure your communications are secure; check your transistor radio and spare batteries. Keep tuned to radio or television news broadcasts. Batten down your house by making sure all doors and windows are securely fastened or strengthened by diagonal strips of heavy adhesive tape. Remove all pictures from walls and all table ornaments; these could be lethal. Anything on your property that could be a wind-borne missile, including rubbish bins, outdoor furniture, dead branches, toys and

tools should be anchored or taken inside. Store drinking water in bathtubs and bottles and check emergency supplies and tools.

When the cyclone comes, remain calm and stay inside. Keep a window open on the side away from the wind and remember if the wind drops suddenly, the calm eye of the cyclone is probably overhead. As the winds will soon resume abruptly with equal violence from another direction, you must shut that window and open another at the far side of the house … if your home is a caravan, you should seek more substantial shelter … if the building shows signs of breaking up, stand in doorways or shelter in the bathroom.

CHAPTER 5

PREPARATIONS

On the afternoon of Christmas Eve 1974, a meeting was called by the executive officer of HMAS *Coonawarra*. Officers and senior sailors discussed what action was to be taken in regard to the cyclone heading toward Darwin.

NOCNA's Cyclone Standing Orders encompassed all the advice given in the cyclone warnings on radio, such as the storage of food and water, the taping of windows, the removal of loose objects around the yard, but also the fuelling of vehicles. At HMAS *Melville* and *Coonawarra* all those actions were carried out.

Tasks were assigned to various officers and senior sailors. Petty Officer Michael French's assigned area of responsibility was Kitchener Street. He was to update all households there on the state of the cyclone and tell them that if they had a problem, they were to attract attention by hanging a sheet or blanket over the railing on their front porch.

During the afternoon, married sailors were given time off to secure their houses and yards while single sailors and Wrans secured the various single quarters, official buildings and grounds.

I was on days off, caught up in my own preparations and looking forward to the dance that night at the Junior Rates Club, known as the Rates. Everybody had their own experiences of the day, and I was fascinated, while researching this book, to hear people describe their accounts of what happened that fateful day.

Most of the houses in Darwin had a similar design. They were high blocked, on stumps, with living areas on the top floor and a laundry, storeroom and parking on the concrete base. Large louvre windows were in every room and the roofs were tin sheets. They weren't sturdy but were delightfully airy in the tropical heat.

Michael and Christine Pilkington's house, and their preparations, were fairly typical. They lived at Wagaman Terrace, Casuarina. Michael, a civilian technician, worked at the Receiving Station at HMAS *Coonawarra*, and Christine and her neighbour ran a creche. The two women were looking after children for very little money. Christine says:

> I just wanted something to do, and I love kids. So, we'd had our Christmas party and then we got together around a few drinks. We'd had an earthquake, a little earth tremor, a few days before and I had kids in the house, and I dragged them all out in case the house fell. So that was that. We had those cyclone warnings and said, 'Oh, here we go again.' Yes, we got a bit complacent.

The Pilkingtons bought their sons a brand-new bike each for Christmas and left them at a mate's place. The boys were six and eight years old. On Christmas Eve, Michael went to pick up the bikes and noticed it was 'raining a bit'. He and Christine hid the bikes in the house to keep them as a surprise for their sons in the morning. Then they kept the boys busy to tire them out, ready for bed.

Liz Grimes was a Wran petty officer, a radio supervisor teletype, based at HMAS *Cerberus* in Victoria. Her parents, two sisters and two nieces lived at Jingili, a northern suburb of Darwin, and Liz came home on leave to celebrate Christmas with them. She arrived in the early hours of Saturday morning, on 14 December, greatly looking forward to seeing her family and some of her Navy friends at HMAS *Coonawarra*. On Sunday, she went to the hockey fields at Allawa to watch the Navy girls play their games.

'It was raining. Not unusual as, after all, it was the wet season,' says Liz.

By Monday, Liz and her family realised that a cyclone was heading toward Darwin, hence the continuous rain.

'On Friday, my family and I cleared our yard of any loose items, locked our peacocks in the pen and got the dogs and cat inside, and filled the bathtub with water,' she recalls.

Liz settled down for the night with her mother and father, her two sisters and her youngest sister's two little girls, one four years old and the other two months old. Their brick home was low blocked, and they were confident it would hold up well to cyclonic winds.

At Naval Headquarters in Darwin, Captain Eric Johnston and his staff reviewed the patterns of previous cyclones. Johnston determined that the patrol boat squadron would probably weather the cyclone better if they were at the cyclone moorings in Frances Bay, off Stokes Hill Wharf, instead of remaining tied up at the wharf. He called all four patrol boat commanders to a meeting at Naval Headquarters at 2 pm, explained the situation and ordered them to recall their crews and sail by 4 pm to the cyclone moorings.

Bob Dagworthy, CO of the patrol boat HMAS *Arrow*, says, 'We were all of the opinion that because of the way that the cyclone was tracking, it was unlikely to pass directly over Darwin. At that stage, we also did not believe it was going to be a major cyclone.'

The plan was for the four patrol boats to come back alongside the wharf early the next morning. The wives and children were to join them there for Christmas breakfast on the wharf beside the boats. NOCNA was going to take charge of the party.

Bob and Christine had listened to the cyclone warnings on the radio and carried out all the recommended preparations, even though they didn't think the cyclone would be a major problem for Darwin. Like other naval families in town, they filled the bath with water, put masking tape on all the louvre windows, removed pictures from the walls and cleared objects from tables, kitchen benches and the sideboard. They lived at 8 Nelson Street, Stuart Park, at the end of a cul-de-sac. In those days, there weren't the tall, established trees that are there now. Bob and Christine have been back a few times in recent years and say the house looks much more protected. They had a beautiful view over the harbour back then, but the house was very exposed, and the block rapidly sloped away to the street below.

After the meeting of patrol boat commanders, Christine drove Bob down to the wharf to join the ship and he said to her, 'Look out in the morning and you'll see us coming back in.'

The chief engineer of the patrol boat HMAS *Attack* (in those days the Navy called them the charge engineer) was away in Melbourne. Bernie Kloss, second engineer on *Attack*, stepped up. He was still a leading hand but now it was his responsibility to keep up the maintenance on all the machinery, air conditioning, refrigeration and generators on the patrol boat. That was a challenging load for a twenty-year-old, but he was capable and got on with the job.

The starboard engine needed servicing and maintenance, which involved pulling parts such as the fuel injectors out of the engine. On 24 December, Kloss was working on the engine parts at the Navy slipway, just around the other side of the Stokes Hill Power Station, on top of a hill at the end of Stokes Hill Wharf. He recalls that at around 2 or 3 pm, two of *Attack*'s able seamen (ABs) came rushing in, yelling, 'Hey Bernie, you've got to get back to the vessel!'

'What's going on?'

'We've got to sail. Urgent. There's a cyclone on the way.'

'Oh, okay,' said Bernie. In that typical Northern Territory attitude of the time, he added, 'Oh yeah, it's just a cyclone. We'll be alright.'

They threw all the engine parts into the back of the ship's vehicle, a Mini Moke, and 'roared back down to the wharf'.

The two ABs came down into the engine room and helped Kloss reassemble the starboard engine. The pressure was on because they had to sail at 4 pm. Even crew members who had no mechanical expertise at all came down to assist.

* * *

Carolyn Spencer, wife of Barry Spencer who was the coxswain on HMAS *Arrow*, had been looking forward to having her mother up from Sydney to celebrate Christmas with them.

'It was to be an exciting time,' she says. 'However, Christmas Eve festivities turned out to be anything but exciting with my husband and the crew from HMAS *Arrow* recalled to the boat to proceed at 4 pm to a cyclone buoy as the winds were intensifying and a cyclone appeared imminent.'

After driving her husband and two other crew members to the wharf, Carolyn returned to her house, one of seven at the 11 Mile, as the wind and rain intensified. Carolyn and Barry's house was the Santa collection point for Christmas morning, so she continued grouping the gifts together for the other households.

Chief Petty Officer Ron Richards had been based in Darwin for fourteen months and worked at the Frances Bay Slipway. The quirkiness of the distribution of naval personnel in Darwin meant he was attached to the Larrakeyah Army Base, was a member of the sergeants mess and lived nearby. On 23 December, he and a chief engine room artificer (ERA), who lived two doors away, went to a cocktail party on the patrol boat HMAS *Arrow* as guests of *Arrow*'s chief ERA.

They heard that a storm was approaching Darwin and the prevailing attitude was 'Okay, so what'. Then they were told it was a cyclone.

At around 3 pm or 3:30 pm the next day, the slipway personnel were advised that, with the cyclone approaching, they would be required to stay at their stations. Richards remembers being told, 'You'll be needed along with your other personnel based at the slipway to stay there to give any possible assistance as required.'

Well, he remembers thinking, *that's great*. Richards was not happy but was resigned. 'It's part of the service job.'

Later in the afternoon, they heard that the cyclone was veering away and were permitted to go home. Not needing to be told twice, off they went. He recalls that 'it would've been about eight or nine o'clock when the wind became very severe'.

* * *

At 4 pm, watchkeeping personnel in the communications station at HMAS *Coonawarra* remained on duty while we off-watch sailors and Wrans continued to secure the bases and prepare for the probable effects of the cyclone. We all had jobs to do whether we were on duty or not, and the base was a hive of activity. We were used to preparing for cyclones and went about the business with practised efficiency. I was keen to get it over and done with so I could enjoy the festivities that evening.

My dear friend Joan Czarnecki was the duty steward at the Wrans quarters, which were almost deserted by dinner time. Many Wrans were away on leave or attending Christmas celebrations at the Junior Rates Club or at other venues around town.

Joan recalls there being 'a few cyclone alerts over the past weeks'. Every time naval personnel had secured property around the base, the cyclone had turned away.

'Tracy was no different,' she says. 'It didn't appear to be so much of a threat. It was Christmas and time to party, and Darwinians knew how to party.'

The duty Wrans officer tasked Joan with securing any object in the Wrans quarters that would be dangerous if it became airborne. Anything over bedding weight was to be put away in the bottom of cupboards or on the floor under our built-in desks. Joan 'went around and did just that. So many things lay around – irons on boards, portable electronics, sewing machines.' Joan also cleaned the bathtubs and filled them with water, which could be boiled for use if the cyclone interrupted the water supply.

* * *

After a mighty effort by Bernie Kloss and other crewmen, the starboard engine on HMAS *Attack* was reassembled in time for the patrol boats to sail to the cyclone moorings. Bernie was still barefoot, covered in grease and wearing a pair of dirty overalls when he went to report 'job done' to the CO. The skipper told him to 'clean yourself up because you've got to get into your Whites and come up to the bridge. We've got to take the vessel out.'

'Excuse me, sir, that's the chief engineer's job,' replied Kloss.

'Yeah, but he's not here so you're it.'

Kloss was not due for promotion to petty officer for another month and working the engines from the bridge was a huge responsibility for a leading hand. He tried to clean the dirt out from underneath his fingernails and found a creased set of Whites down in the locker in the petty officers mess. He'd rarely had the opportunity to wear them. He returned to the bridge: 'The coxswain's there, the skipper's there and we're all dressed in our Whites. I started the main engines, and I was like a rat with a gold tooth. I'm standing up on the flying bridge and I'm king of the mountain now. I'm charge engineer. I'm taking the boat out!'

The four patrol boats went out through the opening between Stokes Hill Wharf and Fort Hill Wharf and proceeded to the three cyclone moorings. Each cyclone mooring consisted of a buoy attached to three seven-tonne concrete weights spread in a triangle on the sea floor. *Attack*, *Arrow* and *Assail* attached to the moorings and *Advance* anchored nearby.

Before a patrol boat approached a cyclone buoy, the anchor would be removed from the anchor chain and secured to the deck. A ship's boat would be lowered to the water with a sailor on board, and the anchor chain would be lowered down to him. The sailor would climb onto the buoy from the ship's boat and attach the chain to the buoy with a securing shackle. The sailor and the boat were then recovered safely back on board the ship. The ship had to remain attached to the cyclone buoy until conditions were safe enough to put a sailor back onto the buoy to break the securing shackle.

'Some patrol boat crew members were given permission to stay ashore,' says Dagworthy, 'but I had a complete operational crew except for my second charge. Everybody else chose to be on board.'

HMAS *Arrow* was able to complete the mooring before dark. Everything was secured and locked away by the time darkness fell and the wind began to slowly increase. Dagworthy 'was completely satisfied that *Arrow* was fully squared away and was ready to ride out the storm safely'.

When *Attack* was finally shackled to the mooring buoy just before dark, they let out two-thirds of their anchor chain and Kloss shut the engines down. The crew had dinner and settled down to rest.

* * *

Like all other naval families, Keith and Kathy Chay had completed their Christmas food shopping and their fridge was stocked with goodies for the following day's festivities. Keith was a warrant officer marine propulsion technician (WOMPT) attached to HMAS *Melville*. His wife, Kathy, had been a leading Wran in the *Coonawarra* Comcentre. She had paid off in January 1974 after completing her four-year engagement with the WRANS.

Their house in Walker Street, Nightcliff, was one of about fifteen Navy married quarters in a small cluster also occupying Rankin Street, Casuarina Drive and Aralia Street. The houses had the standard tropical design of living areas on the upper floor, and laundry and parking on the concrete slab below. All the fences were three-foot-high cyclone mesh with galvanised pipe frames.

Keith had been through a few cyclones before in Darwin, so when he and Kathy heard on the radio about the approach of Tracy, they prepared in the usual way. They took all the necessary precautions such as securing rubbish bins, bikes and other outside items in the laundry and parked their panel van safely under the house. They were ready.

When I put the word out for former naval personnel to tell their stories, Keith sent me an email saying:

> I recall that late in the afternoon the rain had increased, and the wind was getting stronger also. We were in the lounge room looking out the louvre windows along Walker Street in the direction of Aralia Street. We could see the power lines swinging and trees along Casuarina Drive were bending toward the beach. Just before dark, some of the power lines on the corner of Aralia and Walker Streets started sparking as they contacted each other. Then there was contact with the road surface and huge sparking and noise followed by loss of power and lights in the house. No lights to be seen from any of the

nearby houses. I was thinking this is something that will have to be sorted out tomorrow!

* * *

Mick Hamilton was a petty officer electronics technical communications (POETC) at HMAS *Coonawarra*, looking after the crypto for the RERAD – the re-radiation of the fleet broadcast emanating out of NAVCOMMSTA Canberra. Mick and his wife, Doreen, didn't have any children at the time and didn't fancy attending the Christmas celebrations at the senior sailors mess, which was family oriented. They decided instead to join the festivities in the marriage patch at the old transmitter station at *Coonawarra* East, on the 11 Mile, not far from the RAAF Base, with Mick's mate Ray 'Skull' Morse and his wife, Ruth, and another couple, Ned Cooper and his wife, Hannah.

As the winds from Cyclone Tracy grew stronger, the transmitters were falling off the air frequently, and Skull Morse was called in to get them up and running. Hamilton, though off duty, gave him a hand. Doreen, Ruth, Ned and Hannah stayed in one of the residences.

* * *

HMAS *Coonawarra*'s traditional Christmas Eve dance went ahead in the Rates, which was a licensed club with a bar, pool tables and jukebox. I had been looking forward to catching up with Bernie Kloss at the dance and was disappointed that the patrol boats had sailed. I had no concerns for the safety of the patrol boats and their crews. To me, they were invincible. I went to the dance and had a great time, partying with my mates.

Michael Daly, HMAS *Advance*'s new cook, remembers, 'There was a cyclone warning on Christmas Eve so the CO said half the crew could go ashore and the other half would stay on board, and we'd swap over on Christmas Day.'

Other patrol boat crewmen told him, 'It's normal that cyclones come along, and they generally just go past so we've never really thought it was an issue.'

Daly and other off-duty crewmen from *Advance* went to the Christmas Eve dance at *Coonawarra* and he remembers the Rates being 'pretty packed'.

While we were happily partying, Tracy was travelling east-south-east toward Darwin at six km/h. Flash cyclone warnings were being issued by the Bureau of Meteorology advising that the cyclone was expected to make landfall early on Christmas morning. *Coonawarra* and *Melville*'s Passive Defence Orders were implemented, and the bases were fully closed up by 9:30 pm. This meant there were to be no further unessential movements of personnel in or out of the bases. NOCNA Captain Eric Johnston, the Command Operations Officer Lieutenant SM Burrell, and two telephone watchkeepers stayed on duty at Naval Headquarters near the harbour.

The dance at the Rates was also stopped at 9:30 pm. A lot of us were disappointed that we had to finish up early. I'd had a few drinks and would have been happy to party on. But the festivities were over, so off we went.

Daly and his crewmates bought beer to take away with them. *Advance*'s charge engineer lived in a ground-floor unit in a block of flats at Nightcliff, so Daly and his crewmates 'all jumped in his car and drove out to his place'.

They were staggered by the rain that was 'just phenomenal'. When they arrived at the flats and got out of the car, Daly says, 'It was like standing under a hose, the rain was that thick, hard and heavy. It was just torrential. We grabbed our booze and ran into Charlie's unit.'

While we were snug indoors at the Rates, I hadn't realised how bad conditions were outside. A sailor drove a few other Wrans and me back to our quarters. It was only a short distance, about two hundred metres at most, but the rain was heavy and a full gale was blowing, so we appreciated the courtesy. The wind was so strong that I struggled to get out of the Navy van and was stone cold sober by the time I reached my cabin on the first floor. Cyclone Selma had passed by Darwin ten days before and I had expected Tracy to be the same, but this was much worse.

The people of Darwin have been heavily criticised over the years because we didn't seem to take the threat of the cyclone seriously and enjoyed pre-Christmas parties and drinks. We had followed the advice given in the cyclone warning broadcasts and carried out all the essential actions as dictated in the Passive Defence Standing Orders. We had taken as many precautions for our safety as possible. However, nobody could have foreseen the destructive force of Cyclone Tracy when she hit Darwin. Few people had experienced a Category Five cyclone. Many of us had been on the fringes of cyclones and didn't expect – couldn't expect – the truly terrifying ferocity of a Category Five cyclone passing directly overhead.

CHAPTER 6

THE FIRST HALF OF THE CYCLONE

From 9 pm on 24 December, Darwin began to feel significant effects from the eyewall of the cyclone. By 9:30 pm, a full gale was blowing and by midnight, the winds were recorded at 100 km/h. These winds increased with gusts over 200 km/h. The anemometer at the airport collapsed at 5:30 am on Christmas morning after a gust of 217 km/h.

Tracy's path had been erratic. Reports of the cyclone's position and movement were being received from both the Bureau of Meteorology and the RAAF Base, but they did not always agree. Captain Eric Johnston says the resultant plotting 'did not give an accurate picture of the likely movement of the cyclone until late in the day'. We were all scrambling to adjust to a rapidly deteriorating situation.

'Throughout the evening, the wind and rain were worsening,' says Carolyn Spencer, wife of HMAS *Arrow*'s coxswain Barry Spencer. 'I suggested to my mother there was no point in her staying up too late as Christmas Day was going to be full on! I tracked the cyclone from radio cyclone warnings until transmission stopped around 11:30 pm or so.'

Throughout the night, Wrans at *Coonawarra* kept logs as reports came in from the RAAF, the local Bureau of

Meteorology office, radar points, the Australian Broadcasting Commission and ground observers.

In *Cyclone! Christmas in Darwin 1974*, authors Peter Durisch and Robert Howarth note:

> The log of events recorded by the WRANS is considered one of the most reliable accounts of the night when Darwin was destroyed by the most destructive cyclone ever recorded in Australia. The first entry in the log, at 9:30 pm on Christmas Eve, read simply, 'Organisation activated'. Five minutes later the main electricity supply failed. It was restored within two minutes.

The Wrans log noted that fire and repair services were reported as all correct and operational, and the *Coonawarra* station gate was ordered to be left open. At 10:15 pm, Naval Warrant Officer (WO) Maurice Schoer was ordered to the RAAF Cyclone Operation Centre, and through the night he relayed reports of the cyclone's condition to the Wrans at *Coonawarra*. Around midnight, Schoer advised *Coonawarra* that Cyclone Tracy appeared likely to be deflected across to Darwin.

Travel was increasingly dangerous, and people were seeking shelter wherever they could, often in unlikely places. At nineteen minutes past midnight on Christmas Day, a memo noted that two Army personnel were bedded down in the Wrans quarters. Soon afterward, the RAAF reported the eye of the cyclone to be twenty-six kilometres away and stationary.

At the old transmitter station on the 11 Mile, POETC Mick Hamilton was helping his mate AB Ray Morse to keep communications open. I met up with Mick and Doreen in 2020, and he told me about the experience. He said that the authorities in *Coonawarra* contacted him over the order wire (teletype) and told him he was to be in charge as he was the highest-ranking sailor present, and that the Navy's transmitters had to stay on the air, as well as the RAAF's. The RAAF's transmitter station across the highway was a brick building. It was much more substantial and cyclone resistant than the Navy's old, ramshackle wooden structure, which was, as Hamilton said, 'not in real good shape'.

Hamilton was told by the authorities at *Coonawarra* that it was up to him to determine when and if the station had to be evacuated.

There were two ways of contacting *Coonawarra* from the transmitting station. One was a sound-powered ring-down phone like an emergency phone, and the other was the teletype. If the phone contact failed, Hamilton was instructed to advise *Coonawarra* via teletype if the transmitter station had to be evacuated, and then open the hatch of the teletype and pull off the loop current plug. The teletype would then race on an open line, clattering loudly and rapidly, and the operators at the other end would know that Hamilton and Morse had left.

* * *

Able Seaman Michael Daly was sheltering in the ground-floor unit of the charge engineer of HMAS *Advance*. 'They put some candles up and we were just having a quiet drink sitting on the floor in the kitchen,' he says. He knew they were in for trouble when he put his hand on the glass of the front door and could feel the glass pulsing. *Whoa, that's not normal*, he remembers thinking. *I'm not that pissed.*

> I went back and told the others and then a minute or two later, a tree branch came flying in through the glass door. That was it. It was on! We all flew into their bathroom and huddled down. It just howled and howled and howled. At some stage we heard a frantic banging

> on the back door. It was some bloke from upstairs. He came down and someone opened the door for him, and when he entered, he told us that the whole roof was gone. That was the first half of the cyclone before the eye came across. We were all just shitting ourselves.

* * *

'Tracy hit Darwin with a vengeance,' says Liz Grimes. Her whole family huddled together in a bedroom doorway of her parents' house at Jingili, which seemed to them to be the only safe place. They heard louvres breaking, the roof lifting off the house and 'lots and lots of noise'.

* * *

When electricity lines on the corner of Aralia and Walker Streets contacted one another and then the ground, the power went out at Warrant Officer Keith Chay's house in Nightcliff. The wind grew stronger and stronger, and rainwater leaked through the louvres on the windward side of the house. The timber floors were getting wet, and Keith and his wife, Kathy, tried mopping up the rainwater but soon concluded their efforts were futile.

Their dog was distressed with the noise, wind and rain and tried to shelter under furniture. They noticed luminous specks in the water on the floor and realised the particles had come from the ceiling that was slowly disintegrating. Rainwater was leaking through the roof and collecting in the ceiling space. The entire ceiling sagged, and then collapsed onto the floor in the lounge and bedrooms. Keith and Kathy looked up, appalled to see the galvanised iron roofing being blown loose and flying down into the yard. The walls were starting to move, and they decided to evacuate. It was too dark to see if any other houses were also suffering damage, and Keith thought maybe they were just unfortunate to have their house disintegrate around them.

Keith and Kathy planned to head to a small brick building in a nearby tennis court complex. They couldn't go directly over the rear fence toward the building because that was far too dangerous. The wind was coming from that direction, throwing roofing and other debris around.

They picked up their dog, struggled down the stairs and crawled on the ground toward their front gate, keeping as low as possible to avoid flying galvanised iron. Once outside their front gate, they were able to raise themselves up a little and crawl on their knees alongside the fence, which served as a shield, giving them some protection from the debris flying above.

The couple managed to get past their next-door neighbour's house and reached the open gate of the next house along. The householders were sheltering in their downstairs laundry and keeping a lookout with a large torch, when they noticed Keith and Kathy crawling along their fence line. They shone the torch to catch Keith and Kathy's attention and beckoned them in. Keith says, 'There must have been some yelling and screaming but we couldn't hear anything other than the extremely loud cyclonic wind.'

With safety in sight, he and Kathy lowered themselves to their bellies again and crawled toward the torchlight. The householders had only arrived in Darwin a few days earlier and had yet to unpack their removal boxes, which were stacked all around the internal walls of the laundry. The boxes formed a kind of citadel, safe from the elements. The room was already a tight fit for the family, but they welcomed Keith, Kathy and the dog.

* * *

Ron Richards, a chief petty officer attached to the Frances Bay Slipway, was also living in Nightcliff with his then wife and two children. Their house was typical for Darwin, raised on stilts with wall-to-wall louvre windows. Richards 'shut the louvre windows because a bit of water was coming in, but then the wind got so vicious that it blew the rain through the gaps in the louvres even though they were closed'.

Richards followed the advice that had been broadcast so frequently on the radio and television. He paddled through the water, taking his wife and two children into the bathroom for safety as 'the wind was getting stronger and stronger and stronger'.

* * *

Robert Dagworthy, the skipper of HMAS *Arrow*, and his wife, Christine, lived in a Navy house overlooking the harbour. Their home was a basic tropical house on stilts with very wide eaves and louvre windows all around, which Christine describes as 'a lovely house, very airy. We just had overhead fans; no-one had air-conditioning in those days. It was airy and breezy.'

Their dog, Germs, always slept at the top of the stairs near the bedrooms, choosing to sleep as close as possible to the family.

With Bob on *Arrow* and the house prepared for the storm, Christine settled down for the evening. After dinner, she put daughters Siobhan and Kate to bed and then, with the wind and rain strengthening, she put Germs into the storeroom underneath the house. She remembers thinking, *I'll go to bed and try to sleep, and when I wake up all of this will be over.* She was in bed when the louvres on the street side of the bedroom started cracking and snapping despite being taped up. Rain streamed in through the broken louvres.

Christine dressed quickly in jeans and a T-shirt. Thinking she might have to move quickly in the wet conditions, she decided to wear sandshoes instead of sandals. The phone was still working, and the base engineer officer John McCullough rang and said, 'I'm going to come and collect you and the children and take you to a house nearby.'

Bob and Christine had discussed the dangers of flying debris, and Christine had promised him she would not leave the house. However, these conditions were far worse than they had imagined and staying in an exposed house that was breaking apart was far more dangerous than leaving. Christine agreed to go with McCullough and, while waiting for his arrival, prepared the girls and packed a small bag. McCullough came around in a Navy vehicle and parked in the driveway. He crouched low, climbed the stairs and carried the children down, one at a time, and put them into the car. Christine followed, crawling down the stairs, gripping the railings.

McCullough drove them carefully through the heavy rain, howling winds and flying debris to a more sheltered house. Gathered here were the wives and children of other officers of the patrol boats and Naval Headquarters. Christine recalls 'there were maybe six women and about six or seven children inside'. She adds, 'I felt so relieved to be with company. We just sat in the living room and reassured ourselves that we were going to be alright. By then of course it was extremely windy, wild, but there were no windows breaking in this house.'

* * *

Naval Officer Commanding North Australia Area Captain Eric Johnston made his final telephone call to Navy Office, Canberra, at twenty-five minutes past midnight on Christmas Day. He advised the duty staff officer of 'Tracy's imminent presence in the city' and predicted 'severe damage would occur right throughout the city'. By then, part of Naval Headquarters had begun to disintegrate, the trees opposite the building had all been uprooted and 'a car passed down the Esplanade some twenty feet in the air'.

Many years later, he said, 'I well recall the final words from the naval duty staff in Canberra, which seemed somewhat fatuous although well meaning, wishing happy Christmas to myself and my staff. Five minutes later, all phone and radio contact were lost.'

* * *

My dear friend Ruth Turnbull completed her eight-hour watch at the old radio station at HMAS *Coonawarra* at midnight. Taxis were still bringing Navy personnel back to the base, and Ruth decided that if she was going to die, she wanted to be with her boyfriend Chris, a civilian in Darwin. She caught one of the taxis leaving *Coonawarra* to return to the city. 'What used to take ten minutes, took us about four hours,' she recalls.

She told me that when she arrived at Chris's apartment, the unit was awash with water, and she could hear the screech of every nail tearing out of the corrugated iron roof. 'Oh, it was terrible,' she says and for the rest of her life, she 'could not do a thing in the wind and rain'.

* * *

After the dance at the Rates ended, I returned to the new Wrans quarters and settled down in my cabin on the first floor. The wind steadily strengthened and drove the rain horizontally through the gaps in the louvres, spraying water through my cabin. I took shelter in the built-in wardrobe and used a coat hanger to hook the door closed, leaving a small gap to circulate air. The gap was big enough to let a fair amount of rain in too, even though it faced away from the windows. The air was so thick with moisture that breathing was difficult. Surely this couldn't get any worse, I thought.

The fire alarms went off.

Of course they did. *Murphy's Law*, I thought. *Whatever can go wrong, will go wrong.*

We were trained to respond immediately to alarms. We were to investigate the cause, deal with the problem and evacuate if necessary. But a fire could not have broken out with all this water flying around.

I was reluctant to leave the safety of my wardrobe but the alarms were compelling. Shuddering, I clambered out of the wardrobe. The water in my room was already ankle-deep. I waded to the door and emerged into the corridor. The Wrans quarters had a breezeway design and the wind and rain blasted along the corridor from one end to the other. Water raged in rapids to the stairwell at the far end of the building.

I struggled against the wind buffeting through the corridor. I pulled myself along, gripping the floor-to-ceiling metal louvres on the cabin walls. Surely there were no fires. The alarms must have been caused by an electrical fault.

The wind almost toppled me as I crossed to the phone on the far wall. I rang the duty quartermaster gunner and reported the alarms, yelling into the phone. His responses were barely audible over the roaring of the wind. To my relief, he said to ignore the alarms.

I turned around so the wind was now pummelling me from behind. A fierce blast knocked me off my feet and I landed on my back on the deck. The rushing water swept me along the corridor, feet first. I snatched frantically at the metal louvres along the walls, desperately trying to stop. After sliding about ten metres, I finally grabbed hold. A ceiling fitting crashed at my feet, missing me by inches. Panic paralysed me. The torrent of water flowing down the corridor buffeted my body as I gripped the louvres.

I eventually hauled myself up and carefully inched my way back to my cabin. Shivering with shock, I hunched up in the wardrobe again. I was bruised and soaked to the skin, and my hands were bleeding.

* * *

During the evening, *Coonawarra*'s Michael French patrolled his allotted area of the married patch four or five times, but nobody had followed the advice to attach a sheet to the railings by their front door if they needed help. Petty Officer French noted that the wind was so strong that the sheets would have been blown away anyway, but he looked for other signs the occupants needed assistance.

Conditions deteriorated to the extent that French told his wife it was too dangerous for him to proceed out anymore to check on others. He was worried about the safety of his own family, so he, his parents, his wife, their two sons and the family dog took shelter in the bathroom. He says, 'The electricity had gone out at this stage and the wind was howling just like a freight train.' They noticed their neighbours had a light on in their bathroom and that gave them some comfort.

Soon after, they heard 'an almighty crash' as a tall gum tree crashed through the roof over their lounge. More of the roof was torn off by the wind and some of the walls collapsed. French and his family were in grave danger.

* * *

Wran Steward Joan Czarnecki was on duty at the old Wrans quarters at HMAS *Coonawarra*. She remembers the winds getting stronger and stronger as the evening progressed, and the phone system going down. At times, she had to hang on to the wall to steady herself as she made trips between the accommodation upstairs and the sick bay on the ground floor with messages and updates. Wrans dribbled back to the quarters but didn't stay in the same place, so that made it hard to keep track of people.

With their houses disintegrating around them, many residents of the naval housing estate decided to evacuate to the relative safety of the Wrans quarters. The Navy fire truck tried to rescue as many families as possible until it broke down. Joan says:

> The fire truck was used as a wind break to evacuate wives and children from the marriage quarters across the paddock from the old Wrans block. It crawled at a slow pace as the evacuees huddled against its side, until it was as close as it could get to the building, and then we ran out to help get them the rest of the way. The old block recreation room became the marshalling place and it filled fairly quickly.

Smaller cars were blown off the road and damaged. Pregnant women and mothers with children walked and even crawled through the wind and rain from the estate across the paddock to the Wrans quarters. Debris flew around them – galvanised iron, wooden beams, glass and bricks. Aluminium boats were tossed through the air, with one being sighted flying five metres above ground. Another boat crashed through the window of Commander Hall's office.

'Once it became too dangerous to move about,' Joan says, 'I went into the administration office and I shared the small space with the duty officer, a very well-behaved Labrador, a very pregnant woman and a couple of others.' She comments wryly, 'It was cosy.'

The noise of the storm was deafening. There was the terrible roaring of the wind, the screech of metal tearing and blowing away, the thudding of heavy objects such as masonry landing on the ground and the pounding of the rain. If that wasn't enough, Joan watched in awe as the fibro wall shook and pulsed like an off-key wobble board. For years after, Joan felt very uneasy in fibro buildings with tin roofs.

CHAPTER 7

THE EYE AND THE SECOND HALF OF THE CYCLONE

At nineteen minutes past midnight on Christmas morning, while I was curled up in my wardrobe in the Wrans quarters, a phenomenon known as 'explosive deepening' had begun. Tracy's eye had been large, but suddenly the eye started to shrink rapidly. This caused an enormous and progressive increase in the ferocity of the wind within the maximum speed ring, the eyewall of the cyclone.

Most tropical disturbances that reach storm intensity have tornadoes associated with them when they make landfall. The probability that Tracy contained tornadoes is supported by eyewitness accounts that 'twisters' could be seen at some stages, with evidence such as the embedding of a refrigerator in a water tower about fifteen metres high.

Explosive deepening and tornadoes are the explanations for the gross miscalculation of expected wind speeds. There was never any prediction of the reality – winds of more than 300 km/h over parts of Darwin.

At 1:18 am, a message from the port stated that Tracy had slowed to a stop again. That usually results in the intensifying of winds around the eye where winds are already at their strongest. The sudden lull when the eye appears can be unnerving. To go from severe, destructive wind and rain to calm, clear weather in an instant is phenomenal.

Not all of Darwin experienced the eye, and much of the population had no relief from the relentless blasting of wind and rain. However, the eye passed over the northern suburbs of Coconut Grove, Nightcliff, Tiwi, Moil, Wagaman and Nakara, then the airport and finally HMAS *Coonawarra*.

When the eye appeared over Carolyn Spencer's house at the 11 Mile, her neighbour raced across, saying she and her family were evacuating and to follow them to *Coonawarra*. Carolyn, her mother and Mandy, an Old English Sheepdog, struggled down the steps to the car. Carolyn told me that driving to HMAS *Coonawarra* was 'extremely scary with trees down and power lines hanging'. When they arrived, they were directed to the Wrans quarters to shelter on the top floor. Although they were soaked to the skin and cold, at least they felt secure for the time being. Carolyn recalls 'looking out through the louvres and thinking how eerie it was with the stillness of the eye and clouds swirling around'.

* * *

When the winds of the cyclone intensified, civilian technician Michael Pilkington, his wife Christine and their sons Steven and Paul, took shelter in the main bedroom. But the roof peeled off their house, one section at a time. They shifted from their now-exposed bedroom to Steven's bedroom. 'The sound of the roof ripping off was terrible,' says Christine, 'and then the ceiling came down.'

The four of them ended up in Paul's room, the furthest along, and were crammed up against the wall with the water tank above them. Christine adds, 'We had nowhere else to go so we stayed.' She and Michael put the boys in the only shelter available, under the single bed. They were desperately worried about the boys' safety. Years later, Michael told me they couldn't hear anything above the shrieking of the wind, so he had to kick Steven and Paul's legs to get their attention. When the boys kicked back, he knew they were alright. He kicked their feet intermittently through the night and was always greatly relieved when they kicked back.

In the lull when the eye of the cyclone went over, the family evacuated to their neighbour's house. While walking through the yard, Michael found a $50 note. He said, 'Alright, I'll have that.' And he promptly put it in his pocket.

* * *

'At around midnight,' says Warrant Officer Keith Chay, 'the wind died down and the eye was passing over us. Deathly quiet but we knew more was to come yet.' Keith, his wife, Kathy, and their dog were sheltering in the downstairs laundry of a neighbour's house. The two families were crammed into the small space, and Keith 'had considered sheltering in the car just outside the laundry to create a bit more room, but we decided it wasn't as safe as the laundry'.

The wind and rain built up and struck again from the opposite direction. A huge lump of timber crashed through the windscreen of the car. Keith was greatly relieved he had stayed in the laundry. There would have been catastrophic consequences for anyone inside the car.

* * *

Born and raised in Ingham, Queensland, Petty Officer Liz Grimes's father knew exactly what to do during a cyclone. Liz recalls he 'got us to close all the louvres on the windward side of the house and opened all the louvres on the opposite side. After the eye passed over, we reversed the windows.' They believed that probably saved them from total destruction. When the wind and rain hit again, they all resumed their huddle in the bedroom doorway until the cyclone passed over.

* * *

Chief Petty Officer Ron Richards, his wife and two sons were sheltering in the bathroom on the upper floor of their home in Nightcliff. 'The wind was getting stronger and stronger and stronger,' he says, 'and then, the wind just died, stopped.'

The leading edge of the eye of the cyclone had arrived. Richards told me, 'Then it started again, with great ferocity.' He shone his torch toward the end of the house and saw 'the lounge room wall just peel away. It was gone.'

The rest of the house started disintegrating under the onslaught of the cyclone, and then the bathroom wall disappeared. Richards, his wife and two sons were in the bathtub, open to the elements and in deadly danger.

Richards slid his wife and children down the wall to the ground and followed quickly. He gathered them together and put them in comparative shelter under the house, beside the laundry. The family huddled together, wrapped in a blanket, while the wind howled around. Richard recalls, 'With my torch, I could see the debris that was flying around was quite bad. It was unbelievable what was happening. The second half was worse than the first.' His car, a Mazda Capella, had been parked under the house, and, at one stage, its horn started blaring. Richards shone his torch around the side of the laundry and saw the car was being forced up on two wheels against the stumps of the house. When the wind finally eased, the car dropped back down onto four wheels.

* * *

Sheltering in a house near the harbour, Christine Dagworthy was greatly relieved when the eye of the cyclone arrived. She and her two daughters, Siobhan and Kate, were in a Navy house with other wives and children. She says, 'Everything became still. It was eerie. We all heaved a huge sigh of relief.' They just had to wait for the eye to pass and then the wind would come from the opposite direction. They expected the wind to have the same force as before. However, when the eye passed, the wind was so ferocious that the roof was abruptly torn off. The walls shook and broke apart, and timber crashed onto the large Laminex table in the living room. The women scrambled to push the children under the table and just managed to get their own heads under. Christine recalls:

> I had Kate wrapped in one arm and Siobhan in the other. I was on my knees with the girls underneath my body. I had my fingers, my hands, clutching their bodies, holding them tight so they would not be torn away from me. I thought we were going to be killed. I thought there was no way that we were going to survive in these horrendous conditions. That was how we spent the next several hours until it calmed down.

Roofing and walls had collapsed on the table. The debris had the effect of anchoring the table to the floor, giving the women and children shelter and protection from flying building materials.

* * *

At the old transmitting station, Mick Hamilton and Ray Morse worked feverishly to keep the transmitters operational. As the cyclonic winds intensified, the whole building lifted on its foundations. Hamilton and Morse raced around, retuning the transmitters, trying to get them back on the air and keep them functioning. But then sometime around 2:25 am, part of the ceiling collapsed and hit Hamilton on the shoulder. Morse contacted *Coonawarra* advising staff there that Hamilton was injured, and the state of the building was deteriorating. Hamilton told Morse to evacuate everybody to the diesel shed, which was a substantial building with a diesel engine in the middle. That would provide better protection than the houses or the station.

When Morse left, the transmitter station lifted again on its foundations, shook and the roof trusses crashed down and hit Hamilton on the back. Transmitters the size of wardrobes fell off their pedestals, and the building was falling apart. Roof trusses and air-conditioning trunking pinned Hamilton down. He recalls, 'I was buggered, rat shit.'

Despite his injuries, he managed to wriggle out. He couldn't remember how he got out of the building but has a cloudy memory of lying in water in a ditch, being grabbed by Morse and dragged into the diesel shed. Hamilton was in and out of consciousness, lying on the floor next to the diesel engine while his wife, Doreen, desperately tried to keep him awake. He opened his eyes and saw a big fluoro light swinging wildly overhead. After he was helped up, the fluoro fitting snapped and the light crashed to the ground where his head had been.

Mick and Doreen Hamilton, Ray and Ruth Morse, Ned and Hannah Cooper and a black cat continued to shelter in the diesel shed. Adding insult to injury, Hamilton said the terrified cat 'scratched the hell out of me'. At one stage, he remembers 'looking up and seeing these big steel girders holding up the apex of the building and they were swinging in the sky and the roof was gone'.

The transmitting station ceased to operate due to the heavy damage inflicted on communication buildings, antenna systems, feeder lines and other equipment.

* * *

When the leading edge of Tracy's eye reached HMAS *Coonawarra*, Joan Czarnecki recalls, 'All sound suddenly stopped as if someone had flicked a light switch.' It was all 'Go!' as nobody knew how long the calm would last.

The old Wrans quarters had lost its entire top floor. The area was cleared. Joan says, 'A couple of sailors checked upstairs in case anyone was trapped up there under rubble.' All those sheltering in both Wrans quarters were ordered to evacuate immediately and muster in the junior sailors mess down the road.

I clambered out of my wardrobe with relief and joined the crowd heading to the mess. Vehicles collected other residents who had remained in outlying areas and transported them to the mess. We were horrified by the scale of destruction to the base. Joan remembers 'the base ambulance, a Volkswagen Kombi van, lying on its side, and shredded vegetation, twisted metal sheets and bars, demolished buildings'. The sight of all that wreckage was eerie in the calm of the cyclone's eye. Adding to our unease was knowing this was only a temporary reprieve, and at any time we could be hit with the full force of the cyclone from the opposite direction.

Carolyn Spencer, who worked at *Coonawarra* as a civilian admin officer, was concerned about her mother who was suffering from angina pains. She 'saw the executive officer, and he was able to immediately arrange a car and medic to take us to the RAAF Base Hospital'. Carolyn, her mother, their dog Mandy, and a Wran sick bay attendant were driven down the Stuart Highway for a short time until it was too unsafe due to fallen power lines. They changed onto Winnellie Road, which they hoped was an easier route.

* * *

'The noise and the wind began to abate after another half hour or so. Time was going so fast and yet so slow,' says Michael French, who was sheltering with his parents, his wife

Rosemary, their two sons and the family dog in the bathroom of their home. When the leading edge of the cyclone's eye arrived, Petty Officer French decided to evacuate his family from their home and seek better shelter at *Coonawarra*:

> The eye of the cyclone came right over us and I told my parents that we would all go into *Coonawarra* Naval Base. We all went down the back stairs and my mother was quite panicky about the power lines being down. I told her not to worry as we had had no power for some time.

Their car was undrivable, blocked in by debris from their lounge room and a fallen tree. French caught the attention of his next-door neighbour, Maureen Warren, who was backing her car out of her driveway. She stopped for all of them to get in. Maureen already had her mother, her two boys and her dog in the car. French says, 'We added my parents, my wife, our two boys, me and our dog. It was cosy but I felt such relief.'

When Michael French and his party arrived at HMAS *Coonawarra*, they and the other evacuees were directed to the junior sailors mess, where they joined everybody else from the base.

'Once at the mess deck, we prepared for the second half of Tracy's vortex to hit us,' Joan Czarnecki recalls. 'Dining tables were pushed up against both the walls in the dining area, so we could take cover under them as the wind blew out the glass louvres above our heads.'

Michael French adds:

> In the middle of the mess, we arranged tables on their sides in a square formation and all the kids were placed inside this added protection. I had the master keys to the bedding store, which I opened, and all the available pillows, blankets, etc were taken and placed in the square with the kids. We then awaited the next blow.

We were fortunate that the eye took around thirty minutes to pass over *Coonawarra*, giving many of us the opportunity to seek safer shelter and prepare for the cyclone to hit from the opposite direction. After spending hours alone in my wardrobe in the Wrans quarters, I was relieved to be in the company of other Wrans, sailors and families at the mess. There was some distress and anxiety, but I don't remember any panic.

Another Wran and I were in the galley of the junior sailors mess, gathering first-aid gear, when the eye abruptly passed. On hearing the first warning roar of the approaching eyewall of the cyclone, we dropped to the floor and flattened ourselves against cupboards. The windows of the galley exploded inward above us, and glass and debris were blasted across the room. The noise and force of the rain and wind was absolutely terrifying.

I don't know how long we crouched there, too scared to move, but we eventually crawled to the corner of the galley, then made a dash to the servery door and emerged into the comparative safety of the mess. My nerves jangled, and I was unable to settle.

* * *

When the eye finished its transit, Carolyn Spencer was being driven from HMAS *Coonawarra* to the RAAF Base with her mother, a Wran sick bay attendant and her dog Mandy. She says:

> We had to stop and take shelter in the car beside a building. The wind and rain were stronger, the sound of the wind was like a jet plane. Galvanised iron had smashed all the windows in the car, but fortunately we had blankets to protect us, and Mandy lay at my feet. There were two people sheltering in the building and they asked if they could get into the station wagon as

> well, and even though we had six people in the car to weigh it down, it was still rocking.

The rainfall was torrential with winds officially recorded at 217 km/h, although unofficial estimates placed them as high as 300 km/h. Houses and other buildings disintegrated under the onslaught, accompanied by the sounds of flying debris and breaking glass. All services – communications, power, water and sewerage – were severed. The WRANS log of events ends: '0430. All communications lost. Emergency power available. No knowledge of lost.'

The substantial stone buildings of Naval Headquarters, where NOCNA Captain Eric Johnston was working, were almost levelled. NOCNA and his three staff members had a communications link established with his four patrol boats and with Canberra, but this link was lost as the operations room was destroyed at about thirty minutes after midnight on 25 December.

In the first Eric Johnston Lecture, delivered at the State Reference Library of the Northern Territory in Darwin on 7 July 1986, Johnston stated:

> By 0430 the building had been destroyed and my three staff and I were buried. Three of us managed to dig ourselves out and take refuge in the centre cell, which also acted as the Naval Headquarters bar. To my intense delight, not only did I have a packet of cigarettes, which were not touched by water, but the cell bar fridge, although without power, contained some still-cold beer. I say without equivocation that the cigarettes and two beers I consumed are the best I have tasted in my life.

One sailor remained trapped in the headquarters until first light, when NOCNA and the other staff members managed to dig him out unhurt.

CHAPTER 8

THE PATROL BOATS

In the harbour, the four patrol boats battled to survive in mountainous seas and winds reportedly gusting up to 300 km/h.

HMAS *Assail*, HMAS *Arrow* and HMAS *Attack* had been secured to the naval cyclone buoys and HMAS *Advance* had anchored nearby. The patrol boats paid out massive amounts of steel cable to the buoys and used their engines to remain head-on to the storm.

The commanding officer of HMAS *Assail*, Lieutenant Chris Cleveland, noted in his report of proceedings that his ship cast off from the wharf at 1727 (5:27 pm) on Christmas Eve and proceeded to the north cyclone buoy, securing there at 1732 (5:32 pm). The ship and her fittings were then secured for rough weather. As the weather deteriorated, the CO discussed the situation with his charge engineer, who assured him it was not necessary to start the main engines at that stage. They were still on the bridge when the ship broke away from the cyclone buoy at 1:58 am on Christmas Day.

In a January 1975 interview with *Navy News*, Cleveland states, 'We crash started the engines in ten seconds flat and had the second generator going in twelve seconds.' The seas were so mountainous that the side lights of the ship dipped into the sea on several occasions, indicating extreme rolls of around 80 degrees. Cleveland says they 'proved that the main engines don't fall off the mounts past 72 degrees of roll, even if the battery charger does'. Both propellers were frequently out of the water and the echo sounder transducer became airborne many times. Paint was stripped off the radar scanner motor and foremast.

Earlier in the evening, Cleveland had counted nineteen vessels at anchor astern and in close proximity. Visibility had now dropped to less than two metres and there was a real possibility of collisions. He was able to turn *Assail* and place her stern into the sea.

His next concern was for the section of anchor chain which could have been hanging loose after breaking away from the mooring buoy. He dispatched two men to the forecastle to check. The men struggled heroically to traverse the pitching deck in the heaving seas, rain and wind and were able to report that the cable had parted at the hawser pipe and did not pose a risk to the ship. Cleveland was enormously relieved and was able to concentrate on manoeuvring and navigation. He writes:

> At this time, I had not discounted the possibility of capsizing and consequently decided to remain in the harbour. The western side of the harbour provided reasonable sea room and hopefully, few craft. *Assail* ran a racetrack roughly east-west with the sea ahead and astern. No great difficulty was experienced in turning provided that sufficient headway was maintained. Turning at rest was impossible. The ship was easier to control with the wind astern. After running the racetrack for about an hour and a quarter, my confidence in the ship's stability was completely restored and I decided to proceed to sea. At about this time, the wind veered very quickly from ESE to WSW. The departure posed no real problem and once the ship was at sea, I felt relieved and confident that

> she would weather the conditions. She rode very well with wind and sea about thirty degrees on the bow and slow ahead on one engine.

Assail rode out the rest of the cyclone at sea.

* * *

Lieutenant Peter Breeze, CO of HMAS *Advance*, had decided not to tie up on a cyclone mooring and anchored instead in the inner harbour. The wind increased through the night and the anchor started dragging after 1:00 am. Breeze started both engines, weighed anchor and got underway at 1:30 am, attempting to remain headed into the wind.

In his Report of Proceedings, Breeze notes, 'After an hour, no headway could be maintained with accuracy and as a result of the strong ebb tide, the ship was gradually being set onto Emery Point.' He decided to head out of the harbour to sea.

He ordered three-quarter power ahead on the port engine and three-quarter power astern on the starboard engine. Then thirty degrees on the starboard wheel. He was stunned when the ship unexpectedly started swinging to port, something he said shouldn't have happened. There were several more anxious moments when the engines overheated due to air being drawn into the cooling system, but the main engines, radar and gyro compass continued to work through the night.

To compensate for the effect the cyclone had on their forward progress, Breeze steered *Advance* forty degrees off course and managed to guide the ship through that turbulent confusion of waves. The wind howled and the boat heaved and pitched as she met every lumpy wave. Metre by metre, they clawed their way out of the harbour. Breeze noted in his report, 'Once clear, at 0530, the most comfortable courses were steered, and the ship's company cleaned up as much as possible.'

Like *Assail*, *Advance* also rode out the rest of the cyclone at sea.

* * *

As the seas steadily mounted, Leading Seaman Bernie Kloss was having a lie-down on top of his bunk, bracing as the vessel rocked and rolled. He was the second engineer on HMAS *Attack* but had found himself working as acting chief engineer. *Attack*'s commanding officer, Lieutenant Paul de Graaff, called down on the intercom, 'Bernie, Bernie, quick. I think we'd better start those main engines.'

Kloss jumped out, raced down to the engine room and started up the engines. There were normally five people in the engineering branch on the vessel: the chief engineer, the second engineer, a leading hand stoker and two able seamen stokers. *Attack*'s chief engineer was away though, and one of the stokers had been given leave to stay ashore, so there were only three men left, and they were all up and working as the conditions worsened. Kloss, as acting charge engineer, was stationed in the wheelhouse.

In his report of proceedings, de Graaff writes:

> Wind force had steadily increased during the evening and by 0045 on Wednesday 25 December, was such that the anchor winch brake could not hold and about four shackles of cable had been pulled out. Main engines were started, an effort was made to stem the buoy, and the anchor winch was connected up to stop further cable from being paid out. At about 0100 the ship started dragging the cyclone buoy and main engines were used in an effort to remain 1000 yards offshore and to avoid shipping. It was considered dangerous to attempt to heave in the cable and the winch was left connected up.

All around them, boats were breaking loose from their moorings in the mountainous waves and lookouts kept watch to try to avoid collisions. Even so, visibility was poor and steerage so difficult that a barge drifted past, brushing all the way down the starboard side of *Attack*. The CO told Kloss they had to try to keep the strain off the anchor cable as the ship plunged and rolled. Kloss says:

> I had to stay on the throttles doing engine movements all night because it was a manual operation, and the engines were controlled only by hand. There's no seat there or anything, so I'm standing up all night, bracing myself. The coxswain was sitting right next to me in the wheelhouse at the helm, and he was trying to keep a course which was pretty hard. Anyway, there we were. We were going engines ahead. Engines ahead. Stop. Engines ahead. Engines ahead. Stop.

With visibility deteriorating, Kloss says that it got to a point where they didn't really know where the mooring buoy was. 'The next morning, we realised that at different times, the mooring buoy must have actually been behind us,' he recalls. 'We could tell by the anchor chain marks in the paintwork down the side of the vessel that we'd pretty much got ourselves wrapped around it.'

In his report of proceedings, de Graaff writes that the patrol boat was 'dragging the buoy after us' as the ship was thrown around the harbour in mountainous seas. The gyro compass failed, and they had no idea which direction they were facing.

Kloss says, 'The coxswain stood up a couple of times to read the magnetic compass, but it was like hurdy-gurdy. It was just doing a complete rotation one way and a complete rotation the other way because it didn't know where it was.'

Behind the chart table, there was a clock, a barometer and an inclinometer. The inclinometer had a pin on each side at forty-five degrees and indicated how far the ship was leaning over to port or starboard. 'Now, this thing was going like the pendulum on a grandfather clock,' Kloss says. 'It was ticking forty-five degrees one way and forty-five degrees the other way. We were rolling right over, and the seas were breaking on the wheelhouse windows.'

At that stage, de Graaff called for everyone to muster in the wheelhouse or three steps down in the space just outside the wardroom. There were only twelve on board out of a normal complement of twenty. With the ship rolling an estimated eighty degrees, they found it hard to brace themselves. Kloss explains:

> I'm trying to do the engine movements that the skipper's giving me and the coxswain's trying to do the right thing on the helm and there was stuff crashing around the wheelhouse. A tiny little black and white TV broke loose, smashed and slid across the floor. A couple of guys got seasick, and it was getting all slimy on the deck. The smell wasn't exactly like a perfume parlour the next day.

In the darkness and without working compasses, they had difficulty pinpointing the ship's position and direction. When they caught glimpses of the lights of Darwin, they could roughly orientate themselves. Kloss says:

> All of a sudden, we looked across and we could see that the Stokes Hill Power Station, the transformer yard, had taken a big hit from lightning. It was like something out of the Addams Family. There were sparks and lightning and all that kind of stuff and, in a blink, the whole of Darwin blacked out.

In the inky blackness that followed, they could only rely on their 975 Radar. The radar technician, holding on with both hands, continued to give readings to de Graaff. But then at 3:50 am, the wind became so strong that the radar dish on the deck couldn't turn anymore. Kloss continues:

> You've seen radar screens where the centre light goes round and round? Well, the light in the middle of this one just got bigger and bigger and bigger until it was the size of the top of a jam tin and then went, 'Bang!' That was the end of the radar.

Kloss wasn't sure what time the radar failed, but one of the crew noticed the barometer on the chart table had started to drop quickly. Later, Kloss and his crew realised that was when the eye of the cyclone was passing over:

> Of course, I heard stories later that the town itself went very quiet and eerie because you couldn't hear birds and all that kind of stuff. But at sea, out there, the eye didn't make any difference. It was still as rough as it could be. Then we saw the barometer coming back up again and we knew then that we were in for the second onslaught.

They didn't know which way they were facing, where the mooring buoy was, or if there were any other vessels on a collision course. They couldn't get off the mooring buoy because they were shackled to it.

At about 4:10 am, Kloss heard 'a god-awful crash out on the bow. We didn't have a clue what it was. There were no lights out there or anything, and the skipper said the vessel felt a bit different.' No wonder. The anchor chain had been ripped out of the chain locker and had torn the anchor winch off. It then 'flew back, hit the gun turret, twisted the 40/60 Bofors gun around and hit the bridge. If it had gone any further, if we had dipped a little bit more, it would have come through the wheelhouse windows.'

Kloss was still on the throttles with de Graaff, ordering, 'Ahead one engine. Astern the other engine. Stop both. Full ahead both. Full astern both,' while trying to maintain some control of the ship.

They saw a light ahead and, not knowing what it was, determined to keep their distance from it. They manoeuvred astern with the light on their starboard side, then the light was on the port side, then in front of them. 'It got to the point,' Kloss says, 'where we just couldn't get away from this thing.'

> We just kept getting closer and closer and closer, and the skipper gave me the order, 'Half astern both on main engines,' so I went half astern both and we're still getting closer, and he said, 'Right, full astern on them both.' So, I went full astern and we're going flat out backward and next thing, CRASH. The whole vessel shuddered, the windows in the wheelhouse were shuddering and carrying on, but, suddenly, there was no movement. And everybody's holding their breath, and then breathing about twenty times faster than they normally would, hyperventilating.
>
> The time was 4:20 am. One of the crewmen cleared the steam off one of the windows on the starboard side and shined a torch out. He said, 'Hey, boss, we're on the rocks!'

A cheer went up. They were not going to be sunk after all. Being stranded on rocks was an appalling predicament but massively better than sinking in the harbour.

Kloss sent one of the stokers to shut down the main engines and then shut down the generator. The crew scrambled to assess the damage to *Attack*. The engine room, the largest space, was intact, but Kloss remembers damage in every other section.

Attack had run aground close to Doctors Gully, near the Army's Larrakeyah Barracks. She was high above the waterline by the time dawn was breaking. Paul Blanch, the executive officer (XO) of *Attack*, later noted, 'The amount of water pouring off the cliffs and down onto the patrol boat was phenomenal.'

With the sun rising, they were able to see the light they had tried so hard to avoid. It beamed from the masthead of a trawler called the *Clipper Bird*, which had also run aground.

Attack's crew were safe, but the ship had sustained severe damage to the hull, rudders and propellers. Blanch himself had sustained a fractured foot during the pounding.

* * *

The fourth patrol boat, HMAS *Arrow*, broke from her mooring at about 2:45 am. In the pounding waves and ferocious winds, her anchor winch and cable slips were also ripped clean off the deck of the ship.

Arrow had secured to one of the cyclone moorings in the harbour before dark on Christmas Eve. The CO, Lieutenant Robert Dagworthy, stayed on the wheelhouse. 'The command team, the coxswain, the buffer and the charge engineer, all remained in the wheelhouse or on the bridge with me. We had a very reliable committed command team in which the ship's company had full confidence,' he says.

The remainder of the crew, when not required, were down in the mess deck. When conditions deteriorated, waves crashed over the deck, making movements around the ship dangerous, so Dagworthy ordered the crew to come forward. They mustered safely in the forward passageway and the wheelhouse.

Equipment malfunctions occurred early. The life rafts were among the first casualties. Life rafts on *Attack* class patrol boats had hydrostatic releases. If a patrol boat were to sink, the life rafts would be launched once the seawater came into contact with the hydrostatic release. On *Arrow*, when waves first washed over the deck, they activated the hydrostatic releases, and the life rafts were lost overboard. As a result of what occurred during Cyclone Tracy, many lessons were learned in the design, operations and support for patrol boats. At the time, however, nobody could anticipate what was going to happen.

Dagworthy says that the chief engineer had organised the whole engineering team very well. He, or one of the engineers, was always in the engine room, monitoring the propulsion and generation systems. Despite the extreme conditions, Dagworthy thought *Arrow* was riding the storm well, attached to the cyclone buoy. When the eye came, he says:

> I turned to the crew at this time, and I remember I said to them all, 'Well, we're going to survive. The eye is passing, conditions will be the same, but will then gradually reduce. We're going to be okay.'
>
> However, when the eye passed, we were shocked when conditions rapidly became more extreme. Far worse than what we had already experienced. That was when it turned to a disaster for us on board *Arrow*.

The wind and the sea became extremely violent. At about 2:45 am, the ship's motion changed. With minimal visibility and in perilous conditions, Petty Officer Catton crawled forward on the forecastle to inspect the anchor chain. He confirmed that the anchor winch had been ripped out of the deck and the anchor chain was gone. *Arrow* was no longer attached to the cyclone buoy. The ship was underway. Soon after, they developed a problem with the engines.

On patrol boats, seawater was pumped through two inlet points in order to cool the engines. One inlet was partway up the side of the ship and was used in shallow water to avoid sucking up rubbish from the seabed. The other inlet was in the bottom of the vessel's hull. In bad weather, the bottom inlet was used. During Cyclone Tracy, the patrol boats rolled so much that their bottom inlets were exposed and sucked in air. The pumps would not pick up seawater again until somebody opened a small valve higher up from the pump to let that air out. On *Arrow*, an airlock caused the pumps to fail. The engines became overheated and likely to seize.

Dagworthy had considered riding out the cyclone with *Arrow* headed into the wind. With the engines overheating, he realised he would eventually lose the main engines and have no power. The radar had burned out, the gyro had toppled, and visibility was negligible. Dagworthy says:

> The sea spray was such that you couldn't see. When I went to the flying bridge, the wind was driving so hard that I felt the sea spray could damage my eyeballs. I even tried using a diving mask to no avail. The wind, the rain and the spray were so intense that we were completely blind. I could not even see the bow of the ship from the wheelhouse. The noise was also so extreme that I had to shout orders in the ears of the helmsman and the engineer.

He conferred with his command team and decided to attempt to run the ship into the mangroves on the far side of the old slipway at Frances Bay. He edged the vessel around, trying to be conservative with the main engines. While he was gingerly manoeuvring, a wave hit and washed *Arrow* sideways down onto Stokes Hill Wharf. Relentless waves hammered the ship against the pylons of the wharf.

With the engines threatening to fail at any time, Dagworthy 'made the frightful decision to abandon ship'.

Every time a wave smashed against the patrol boat, the vessel would rise to be almost level with Stokes Hill Wharf. When the wave dissipated into a trough, the ship would drop down below the wharf. With the wind shrieking around their ears, Dagworthy ordered the crew to get onto the wharf whenever the ship rose high enough. He stayed in the wheelhouse, prepared to manoeuvre with whatever engine power was left.

Dagworthy says the XO, Sub Lieutenant John Jacobi, and the buffer, Petty Officer Les Catton, went onto the deck and took charge of the crew jumping from the ship onto the wharf, whenever it was possible to do so. 'They did a great job,' he adds. 'Able Seaman "Mac" McLeod in his support role was heroic during this evacuation.'

Most of the crew scrambled onto the wharf by climbing onto *Arrow*'s funnel casing, then onto the portside guardrail and onto the wharf safety rail. Every second, they were exerting their utmost strength. Jacobi searched along the ship's port side for the safest escape route for the crew. But while attempting to reach the wharf himself, he fell into the sea. The electrician, Kevin Rainbow, missed the wharf and fell back on the deck, seriously injuring his legs. Jack Rennie was also reported to have missed the wharf and fell between it and the ship.

Les Catton made it onto the wharf, but crewmates saw him hit by flying debris and knocked back into the harbour.

The ship took in more and more water and became too heavy to rise to the level of the wharf. Dagworthy told the sailors remaining on board, 'We have to fully inflate our life jackets, take our chances and leap into the harbour.' He did not think he would survive. 'I was washed beneath the wharf and forced by waves through the wharf. I was smacked around.'

Arrow was pounded repeatedly against the pylons, driven under the wharf and was subsequently sunk.

Dagworthy was later found on the foreshore by other *Arrow* survivors. He had some injuries and was in a state of shock. Sub Lieutenant John Jacobi washed up ashore below the wharf gatekeeper's hut. Kevin Rainbow was found injured but conscious on mud flats. Many of the crew carried injuries and most were suffering from shock and exposure.

Able Seaman Robert (Mac) McLeod was later awarded an Australian Bravery Medal for assisting his fellow crew members to safety.

CHAPTER 9

CHRISTMAS MORNING

The width of the area of Cyclone Tracy's destructive winds is estimated to have been less than forty kilometres. But even so, she was a very intense cyclone. Exacerbating factors such as explosive deepening, tornadoes and severe lightning weren't anticipated and couldn't be withstood.

In addition, the eye moved slowly at about eight km/h, so Darwin experienced destructive winds for about five hours. It was estimated that approximately 280 mm of rain fell over Darwin in twenty-four hours.

By 6:30 am on Christmas Day, the winds were abating. Cyclone Tracy crossed the Adelaide River at about midday, degenerated into a rain depression and then moved south-east across southern Arnhem Land into the gulf country of Queensland.

Only 408 buildings out of a total of 10,000 remained intact. Darwin was left without light, power, water and sanitation.

In his lecture 'Operation Navy Help', Eric Johnston, Naval Officer Commanding North Australia Area, described the results of Cyclone Tracy. He said, 'On the harbour, the scene had to be experienced to be believed. At the height of the cyclone just before and just after the eye passed over Darwin, the best description can be to liken the harbour to a maelstrom.'

He added, 'The scene at first light was beyond belief – the harbour empty, every building within eyesight destroyed with the exception of Government House, and shortly afterward the survivors of the *Arrow* arrived outside the wrecked headquarters to report their loss.'

Bob Dagworthy, CO of *Arrow*, says, 'Some of the men were in very bad condition and had to be taken to hospital. In particular, Kevin Rainbow, the electrician, had serious lacerations. The navigating officer, Sub Lieutenant Andrew Birtchnell, was also found. He was suffering from hypothermia and taken to hospital.'

Jack Rennie had been killed after trying to abandon ship when *Arrow* had struck the wharf. His body was found by the coxswain Barry Spencer that morning. Les Catton was still missing.

The Navy also lost a workboat and a large fuel lighter, which was later recovered. The civilian shipping damage was extensive as well. According to the harbourmaster's report, 'At least twenty-nine vessels were sunk or wrecked.' These included several large prawn trawlers that were based in the port, a passenger ferry called the *Darwin Princess*, and the three-masted steel schooner *Booya*. In all, seventeen people were officially reported to have died or were listed as lost at sea, presumed dead.

* * *

HMAS *Attack* had beached under the cliffs at Larrakeyah. The ship's company were in ignorance of the fate of *Arrow* and her crew, and of the severity of the damage caused to Darwin. A naval party searching for survivors from *Arrow* eventually met up with members of the crew of *Attack*, but in the meantime, Acting Charge Engineer Bernie Kloss says:

> So now, we can see around the harbour, and we can see trawler hulls, vessel hulls, some in the mud, some in all

> kinds of trouble, but we were under a cliff just at the base of Larrakeyah, the Army barracks at the time. It was a little bit further around than Doctors Gully because we couldn't see the town. We could just see, if we got right down the stern, and tried to look around the corner, we could just see the Holiday Inn on the Esplanade. But that was basically it.

With *Attack* stabilised but stuck fast high on the rocks, what to do next? The CO, the XO and two petty officers decided to check on their wives and report to Naval Headquarters. As CO Paul de Graaff was about to climb down the rope ladder to the rocks, Kloss said he looked around at him and said, 'Bernie, you're in charge.'

Kloss looked around, thinking, *In charge of what? There's not much here.*

The crew watched the four men in their white uniforms scramble up to the top of the cliffs. When they were out of sight, the rest of the crew looked to Kloss and asked, 'What are we going to do?' Using his initiative and unexpected authority, Kloss detailed two sailors to salvage a table that had washed up on the beach and set it up on the sand with rocks propping up a missing leg. He detailed two others to stack up the beer in the wardroom into milk crates. Mouths open, they objected, 'We can't do that. That's the wardroom beer.'

Kloss replied, 'Well, what are they going to do to us?' After all they'd endured, getting into trouble with the officers was the least of their worries.

'Are you sure?' they asked.

'Look, I told you to do it. The skipper told me I'm in charge. It'll be on my back, and I don't really care,' Kloss replied.

'Righto,' the sailors said and set to with a will. They emptied the cans of beer into the milk crates and lowered them down to the ground on a rope.

The cook had made the crew a big batch of food for Christmas lunch, and Kloss told two other sailors, 'See if you can go in and salvage the tucker.'

Another bucket was promptly filled with food, lowered down to the rocks and set out on the table. Kloss remembers:

> There were chicken wings with a bit of milk spilt on them, prawns with a bit of tomato sauce or something smeared on them. Anyway, we tucked in. We had a few beers, had a bit of tucker, and after about three beers, that Dutch bravado kicked in again – 'yeah, we could do that again.' That was alright. It wasn't too bad. Here we are, we're alive, and we're having a few beers.

The rest of the crew agreed, 'Yeah, yeah, Bernie, that's alright. We'll do it next time a lot better than that.'

Kloss said, 'Yeah, yeah, righto' – but he fervently hoped there wasn't going to be a next time.

* * *

HMAS *Coonawarra* was in a sorry state. The old Wrans quarters lost the entire first floor. Ninety per cent of the houses on the marriage patch were severely damaged or destroyed. The communications receiving station where I had worked was destroyed, as was the old transmitting station.

Having spent the first half of the cyclone in the old Wrans quarters and the second half in the junior sailors mess, Joan Czarnecki remembers:

> When the howling finally stopped around 6 am on Christmas Day, the scene outside was eerie. The grey sky, the heavy mist hanging in the air, the destruction and not a sound, no crickets, not even a bird calling. The first loud sound after that came from a lone aircraft flying overhead.

> Not long after, a sailor walked into the mess, blurry-eyed and asked what happened. We said, 'Tracy.' He had come back very inebriated early in the night and had fallen asleep in his bunk and slept through the whole thing, miraculously unharmed by bits falling around him. Not all were as fortunate.

Incredibly, I know of three other people who slept through the cyclone: a Wran, a soldier and a civilian. They were all in very secure buildings, and sleeping pills or alcohol kept them in deep sleeps. Two of them told me about waking up to find the power off, being unable to make a cuppa, and of their shock and distress at looking out their windows and seeing utter destruction. They were spared the terror of enduring the cyclone, but nothing else. Their dedication to the clean-up was exemplary.

Many of us at *Coonawarra* immediately set about clearing away debris in the junior sailors mess, the Rates and galley, while patrols were sent out to find survivors among the naval population. Most of us spent the morning of Christmas Day looking for our friends, checking the damage to the base and picking up debris. The destruction was horrendous. Cars were flung on their sides. Roofing iron, walls, furniture, other debris and personal belongings littered the ground. The task of cleaning up was overwhelmingly huge. I, for one, didn't know where to start, and wandered around for a short while, trying to take in the scale of the destruction. I remember picking up a piece of debris and then wondering what on earth to do with it.

Some buildings at HMAS *Coonawarra* still had power, perhaps through generators.

The devastation of the houses on the marriage patch was alarming. The front stairs were all that was left of one sailor's house. Fortunately, he and his family were away on leave.

When we were allowed to return briefly to the Wrans quarters, I discovered all the clothes and shoes in my wardrobe were soaking wet and likely to be spoilt by mould very soon. Fortunately, I'd put my camera and photo album in a plastic bag while we were preparing for the cyclone, and they were alright. I didn't even think of taking photos though.

My dear friend Dot Wright and I had bought each other Parker pens for Christmas. They were still wrapped. The paper was wet and ruined, but the pens were undamaged inside their boxes. We exchanged them, the only concession to Christmas that day.

Another Wran told me that a man was in my car, parked outside the quarters. During the cyclone, my pride and joy, a green Toyota Corolla, had been lifted and smashed against the car parked beside it. A few windows had been broken. Sure enough, a sailor was lying in an exhausted sleep among the broken glass on the back seat. My car had been the closest shelter he could find. It must have been terrifying for him.

While we were clearing away debris, we were misinformed that HMAS *Attack* was sunk and two sailors killed. The thought that Bernie Kloss had been one of them was devastating. Dot and I were friends with a few of the sailors on *Attack* and were heartsick at the loss of the ship and crewmen.

Later, we were told that *Arrow*, not *Attack*, had sunk and that two of her sailors had indeed been killed. Relief at the safety of *Attack* and her crew was mixed with terrible grief at the loss of *Arrow* and two of her crewmen.

* * *

Petty Officer Michael French had spent the second half of the cyclone in the junior sailors mess, helping to protect evacuees. When the wind and rain abated in the morning, he, his wife Rosemary, their two sons and Michael's parents 'moved up to the senior sailors mess, which had suffered relatively little damage'. At around 8:30 am that morning, the supply officer detailed French to drive into Darwin to check out Navy assets there and see what state they were in. French remembers, 'That day, I also went over to RAAF Darwin to set up a relationship in regard to finding out who to talk to. That night was the earliest night I was going to have for a long time.'

* * *

Civilian technician Michael Pilkington's neighbourhood had been devastated. While taking in the utter destruction of their neighbourhood, Michael's neighbour asked, 'What have you got in your freezer?' Michael and Christine had just restocked their freezer, ready for the Christmas holidays, so it was full, but the power was out and the food would not keep for long. They all feasted on prawns and oysters, which were expensive and would have been the first things to spoil.

Young Steven and Paul were resilient. The ceiling had fallen in on their new bicycles, but the bikes were undamaged. The young lads went riding through the debris in the streets.

Christine says, 'We didn't expect it to be that bad. We'd had cyclones before, and we had cyclones after, but nothing like that.'

* * *

Warrant Officer Keith Chay and his wife Kathy, along with their dog, had waited out the cyclone in a nearby laundry with their neighbours. It had been a tight squeeze, but they were safe and sat it out until the wind and rain eased at around first light. Keith says:

> Coming outside to inspect the damage was something to be remembered forever. It wasn't just one or two houses which were damaged. It was an entire street of an entire suburb of an entire city! Looking back

> down Walker Street to our house, all that was left was a kitchen wall and part of the lounge wall. Roof sheets and other debris were in every yard and every street. Steel power poles were warped and twisted. Every house was completely wrecked.

When their house collapsed around them during the cyclone, they had planned to evacuate to a brick building at the tennis courts behind their house but found refuge in a neighbouring house instead. Just as well, because the brick building was demolished.

Keith and Kathy walked around the other houses in the married quarters area, checking on the occupants. This was a daunting experience. Their neighbours across the road were in their front yard. Keith says, 'As I approached them, I could see things were not good.'

The family's eldest boy, a five-year-old, had been crushed beneath his bed and had not survived. This was 'a tragedy none of us were equipped to deal with', recalls Keith. They provided whatever comfort they could.

When Keith and Kathy returned to their own yard, they searched through the rubble for their belongings. In particular, they looked for their photos and wedding rings, without success. Their panel van was still intact and drivable, but the rear doors had been grit-blasted back to bare metal.

Throughout the morning, various rescue groups attending the area included staff from HMAS *Coonawarra*. Their advice was to leave the neighbourhood and come to the base, which, being further inland, had not suffered as badly as the Nightcliff area.

Keith and Kathy Chay packed up their panel van with their few remaining belongings and their dog, scrounged some fuel from damaged cars and headed out to *Coonawarra*.

* * *

When the eye passed over his home in Nightcliff, Chief Petty Officer Ron Richards and his family moved from their bathroom to the underside of the house and huddled together in a blanket until first light. His wife's leg was gashed. They were only in their night wear and Richards searched the debris for clothes for them. He found one of his old coats and his uniform hat, but that was all. His wife and children were badly shaken after spending the night so exposed to the storm, and Richards, himself shaken, wanted to get them to help.

The car's footwell was full of water but Richards removed the manufacturer's plugs on the floor and drained the water. He found his car keys and was surprised and relieved when the car started on the second try. He got his wife and two sons into the car, and the family drove to a church hall in Nightcliff, but it was full of people.

He decided that *Coonawarra* would be better and wheeled his way up Bagot Road toward the naval base.

'The landscape was totally devastated,' he says. 'It was a road that I used to drive up regularly, back and forward to *Coonawarra* and up to Darwin, on our way down to Howard Springs and places like that, but it was now unrecognisable.'

When they finally arrived at *Coonawarra*, Richards says, 'A very nice chap at the gangway motioned me inside and asked who I was.'

Richards identified himself and was told, 'Thank you very much. Could you please go around the corner?' There they were met by another greeter who directed them to one of the accommodation blocks.

'I don't know which one it was,' Ron muses. 'There were multi-storey accommodation blocks, and we were taken into a room with a bed and so forth, where we collapsed.'

A short time later, a young lady he assumed was a Wran knocked on the door and said, 'Chief, can you please bring your family down to the cafeteria.'

Richards replied, 'Gladly,' because they'd had nothing to eat since dinner the night before. They had a very welcome meal and cups of tea and coffee and were then directed back to their room.

* * *

In the suburb of Jingili, Petty Officer Liz Grimes had spent the night with her parents, two sisters and her youngest sister's two little girls, one four years old and the other two months old. They were all uninjured, but her parents' house had sustained considerable damage. Liz says:

> In the morning, after Tracy moved on, we opened the front door and saw the devastation in our street. It certainly was a different place. Power poles (made of steel) were completely bent, a lot at forty-five-degree angles, power lines were everywhere, the coconut tree was uprooted, and cars damaged. My mother's VW had a dent in the roof big enough for a baby's bath. Dad's van had the engine cover missing, and one tyre slashed, and one side of the van was extensively scratched from flying roofing.

She also recalls, more humorously, opening their front door and seeing their cat Ming come out from wherever he'd been hiding during the cyclone. 'He walked,' she says, 'to the front door, looked around at the devastation, turned around and walked back inside. The whole family had a little chuckle and that seemed to help relieve a little tension. We were all thankful we survived with no injuries.'

Their house was at ground level and water damage made it uninhabitable. The family went across to Jingili Primary School and found a place to sleep at night, while working at the house during the day.

* * *

Carolyn Spencer, her mother, a Wran sick bay attendant and Navy driver had been driving to the RAAF Base Hospital from *Coonawarra* during the eye of the cyclone when the far eye wall hit them. They had parked beside a building, taking in two other people needing shelter. 'We sat it out until daybreak,' says Carolyn.

They had a dreadful shock when the devastation became clear. The building they had sheltered alongside had been demolished except for one remaining wall – the one they had parked beside. They'd had a lucky escape.

As they drove toward the RAAF Base, they could see that houses were destroyed, power poles were twisted and electrical cables were hanging loosely.

Carolyn says, 'At no time did I ever consider my husband and the *Arrow* crew would be in danger although my thoughts were on him, and I couldn't wait to tell him of our experience.'

* * *

Several Navy wives and children, including Christine Dagworthy, had sheltered together under a Laminex dining table for hours until the cyclonic winds abated. None of them were injured; however, Christine says, 'The next day you could see where my fingers had bruised the girls' stomachs, where I'd held them so tightly.'

They went downstairs and saw that 'everything was devastated. We were in shock, and we all just cuddled our children. I couldn't think or do anything else. I didn't want to move. I just didn't want to leave the property or move from where I was.'

Even though Kate was only eight months old, she stayed very quiet as Christine cradled her. Two-year-old Siobhan cuddled close to her mother. 'There was devastation all around.

Absolute devastation,' she says. 'Every house was destroyed. Every person quiet. We just waited. There was nothing any of us could do. It was too dangerous to move.'

Sometime later, around mid-morning, people from the Navy came to see how they had fared and broke the shattering news that the *Arrow* had sunk. The women and children were taken to the Larrakeyah Army Base.

* * *

When the storm had increased in intensity on Christmas Eve, Navy cook Michael Daly and a couple of other crew members had taken shelter in the ground-floor unit of *Advance*'s charge, along with his family. Daly says:

> When the eye arrived, it was dead quiet. On the radio all week they tell you what to do if a cyclone actually hit the land, so we all knew to go into the other side of the unit, which was the main bedroom, and we got under the bed. There was the charge, his wife, their daughter, and about three of us. When we were under the bed she started to roar again. I sort of lost track of time, and we stayed there until it got light in the morning, and it was all quiet. We went outside and went 'whoa'. It was absolute devastation.

Daly and his mate Ray Minette, an able seaman electrician (ABET) on *Advance*, went from house to house, helping people. In one place, an old lady was unable to get out of the bathtub in which she had been sheltering. Her children were also ageing and couldn't lift her, so Daly and Minette stepped in, got the old lady out of the bathtub and helped make her comfortable.

Daly says, 'Every house we went to, someone proffered either a bottle or glass of spirits to us and said, "Get this into you, boys." By the time we'd done half the street, we were half sloshed. But that's what everyone was doing. That's what you did.'

Tragically, Daly and Minette came across the bodies of a couple of elderly people. The two sailors rolled the corpses up in a carpet because they didn't know what else to do and had to leave the bodies where they lay.

They finished checking the rest of the street and then realised, 'Shit, we'd better get back to the wharf and find out how the boat fared.' They decided to head to *Coonawarra*, thinking someone there would surely know where the patrol boats were and how they'd weathered the cyclone.

Daly and Minette borrowed an open utility. One of the survivors of the cyclone was coughing up dark blood. Something was clearly wrong with his lungs, so the sailors decided to drive him to hospital first and lifted the injured man into the back of the ute. The drive to the hospital was difficult and painfully slow because, as Daly says, 'All the power poles were blown over. They were all sort of folded, like a neat fold right at the bottom, and they were all across the road, so it took quite a long time.'

They finally arrived at the hospital. 'You couldn't get in the front door – it was packed with people.' Daly and Minette unloaded the injured man from the utility, carried him in and 'plonked him inside somewhere and told someone about him'.

One of the side doors of the hospital was open and Daly saw a nurse using a squeegee mop to swish blood and water out of the door. He remarks, 'One can only imagine how many injuries would have been being treated in there. Anybody who wasn't in shelter during the night were either dead or nearly cut to pieces because of the amount of tin that was flying around from all the tin roofs.'

The two sailors then got a lift to *Coonawarra* but, unable to find out how the patrol boats had fared, they went down to the wharves. They soon found HMAS *Attack* washed up on the rocks and tilted on her side. Daly and Minette returned to *Coonawarra* and ran into Barry Spencer, the coxswain on *Arrow*, and another crew member, who confirmed that *Arrow* had sunk.

* * *

HMAS *Attack* was on the rocks below the cliffs at Larrakeyah, and it wasn't going anywhere. Leading Seaman Bernie Kloss, acting chief engineer, had been left in charge when the commanding officer, the first officer, the coxswain and the buffer had gone ashore to check on their families and report to Naval Headquarters. When they returned, clambering back down the cliffs, the officers were unimpressed.

Kloss and the rest of the crew had no idea what was going on in the town and were, Kloss says, 'sitting back on the rocks and the sand with our feet up' in what looked 'like a backpacker's backyard. There were chicken bones and empty beer cans all over the place.' They were tired, shaken and 'were in a pretty fine old mood'.

'What's going on?' the CO demanded.

Kloss thought, *We were nearly dead a couple of hours ago so all this doesn't really matter.*

'Merry Christmas, boss,' Kloss replied. 'The beers still might be cold enough to have a cool drink, a bit of chicken there.'

The CO sighed. 'Alright. I suppose. There's nothing else we can do right now.' He sat down and had a couple of beers and a bit of chicken too.

The returning men told the crew that there was nothing left of Darwin, no houses, people were dead, and so on. Kloss thought that was rubbish. He thought they were exaggerating and there might only be a couple of houses damaged. He and one of the other sailors climbed to the top of the cliffs to see for themselves:

> We came out on the top and just looked around Larrakeyah and we couldn't believe it. The houses didn't look like houses anymore because everything was gone. Just floors and stumps. We were just standing there with our mouths open for a while. I had a Super 8 movie camera, and I've actually got footage of the exercise we did before Christmas Eve, and I've got footage of the boat up on the rocks, and other things.

* * *

Able Seaman Geoffrey Stephenson had been on duty at the old radio station at *Coonawarra* during the cyclone. He was worried about his wife and two children in their house in the northern suburbs but had to stay at his post. When released from duty on Christmas Day, he drove through total devastation from the base toward his house. Stephenson feared, quite rightly, that his house must have been destroyed.

He got within 200 yards of where his house had stood when the road became impassable. He then had to walk the rest of the way. Stephenson made an initial check of the wreckage but to his relief, he couldn't find any trace of his family. Hopefully, they had evacuated safely. He began to worry again though, returned to his house and did a systematic search of the wreckage. After four hours, he found the bodies of his wife and two children in what was left of the main bedroom. They appeared to have been sheltering under the double bed, which had collapsed during the cyclone and crushed them.

The loss of Geoff's wife Cherry, six-month-old daughter Kylie, and four-year-old son Kenneth affected us all. The tragedy was compounded by the death of the youngest son of Able Seaman Bruhn. I couldn't process that. Not children. Not the death of children.

CHAPTER 10

CHRISTMAS AFTERNOON

HMAS *Assail* rode out the cyclone at sea. At 7 am, her commanding officer Lieutenant Chris Cleveland decided that the weather had abated enough for *Assail* to attempt to re-enter the harbour. The ship steamed past the wharf at 9 am and the crew, awed, took in the destruction of the area. The CO wisely decided against attempting to berth at the wharf, instead anchoring in the lee of Talc Head at 9:30 am.

At 12:37 pm, *Assail* weighed anchor and proceeded to the inner Stokes Hill Wharf. They secured port side to the wharf, immediately forward of what used to be the Patrol Boat Landing.

Assail had not suffered any major structural damage during the night. Relatively minor damage was caused to the port shaft and propeller, which had been fouled by an eight-inch hawser. A forty-four-gallon drum of lube oil, which had been securely lashed beside the boat cradle, had broken away and cut loose the port life raft that was missing. Both lifebuoys were washed out of their bridge stowage.

In his report of proceedings, Lieutenant Cleveland says the ship's company's 'conduct throughout the cyclone was magnificent and they performed their duties to the highest standards and traditions of the Naval service'.

* * *

Like *Assail*, HMAS *Advance* had left harbour and rode out the cyclone at sea. *Advance* suffered heavy damage to her shafts, propellers and 'P' brackets, but had weathered the cyclone. Her CO, Lieutenant Peter Breeze, says in his report of proceedings, 'When the seas had abated sufficiently, course was reversed and *Advance* proceeded into Darwin Harbour, the devastated city being viewed with shocked amazement by those on board.'

After a survey of the wharf area, *Advance* followed *Assail* to the berth at Stokes Hill Wharf at 1:20 pm.

Breeze reported, 'The damage to *Advance* consisted of a bent and holed starboard propeller (cause unknown), damaged rocket flare locker, bent stanchions and a missing life raft. Much paintwork was stripped off the superstructure by the estimated 200 knot winds.' He added, 'Those on board during the cyclone proved very conscientious and capable when subjected to extremely adverse conditions.'

Advance's cook, Michael Daly, and other members of the crew who had stayed ashore on Christmas Eve, waited at HMAS *Coonawarra* for news of *Advance* and *Assail*. Daly 'helped the chefs do some cooking and all that sort of stuff – we did what we could do and waited until we got some news about where the boats were'. When they heard that *Advance* and *Assail* had tied up at the wharf, the sailors made their way back to their boats. Daly says:

> *Advance* looked pretty bad when we actually got back on board. I was overwhelmed when I saw my galley. It was upside down. Everything had come out of the cupboards and the fridge. The deck was awash. I was, however, very glad to see that everyone on board was okay apart from being exceptionally weary and shocked from the experiences they had just been through. The stories abounded, one of which was that when they tried to get out of Darwin Harbour, there was the odd boat or yacht

> flying across and over in front of them. They had a really scary and tough time of it.

Daly got stuck in and cleaned up the galley, then checked the freezer below and the dry store for food. Usable stocks were low, so the married men went home to see their families and to bring any food they had back on board.

* * *

Carolyn Spencer, her mother, their dog Mandy, a Wran sick bay attendant (SBA) and their Navy driver finally arrived at the RAAF Base. Carolyn's mother had rallied, her heart condition settling somewhat, and the two women were dropped off to shelter in the prison at the RAAF Base. The Wran SBA checked to see if she was needed to assist the RAAF personnel. When Carolyn and her mother drove back to *Coonawarra* mid-morning, they were told that all dependants were to muster in the junior sailors mess to get something to eat and drink. They hadn't heard from Carolyn's husband, Barry Spencer, the coxswain on HMAS *Arrow*. Carolyn says:

> As we were heading that way, an officer approached me saying HMAS *Arrow* had sunk with the loss of all lives. I couldn't believe what I was hearing … I raced to my car to drive to Darwin Harbour; however, my car wouldn't start with so much water in the engine. One of the Wrans, who I worked with, kindly offered me her car. However, I realised it probably would have been unsafe to drive so I went back to the cafeteria and weighed up my options as to what to do next.
>
> What seemed like hours passed and realisation was setting in. Suddenly, I looked up to see a bedraggled-looking bloke, wet through, wearing shorts, runners and a foul weather jacket heading my way. It was Barry!

I completely lost it! Barry had taken some of his crew to the hospital with cuts and abrasions. He told us that *Arrow* slammed into Stokes Hill Wharf and that he made sure all crew members were accounted for, either going over the side or onto the wharf. He then went on to say that his POQMG had fallen through the wharf, and he found his AB washed ashore at daybreak.

* * *

When the winds died down, Petty Officer Mick Hamilton went to the sick bay at HMAS *Coonawarra.* His back had been injured when the roof of the transmitter station at the 11 Mile collapsed on him. 'I was hunched over like Magilla the Gorilla,' he says. 'I was humped over and really in pain and for a couple of days they were just giving me Digesics.'

Hamilton and his wife, Doreen, are grateful they were at the old transmitter station for Christmas. Their unit in Casuarina had been completely destroyed in the cyclone. All Doreen found in the rubble was a frying pan and their wedding album.

Hamilton was good friends with a fellow sailor, Geoff Stephenson, who lived two doors down. They used to moonlight, pumping petrol at the Shell Garage at Trower Road. Mick and Doreen were deeply distressed at the news that Geoff's wife, daughter and son had been killed during the cyclone while Geoff was away from them, on duty.

* * *

Wran Radio Operator Ruth Turnbull had a particularly interesting time getting back to HMAS *Coonawarra.* A friend of hers, Bede Eresman, a photographer for *NT News*, 'was driving around taking photos', and Ruth caught a lift with him. He stopped to take pictures of 'the big crack in the

Travelodge' and 'two cars in the Travelodge pool'. Lieutenant Brian Swan from HMAS *Melbourne* (R21) later recalled the 'small sedan that had been picked up by Tracy' and 'thrown against a fourth-floor brick wall', before ending up in the Travelodge pool. He knew it had been thrown against the wall by the 'number plate that was stuck in fourth-floor brickwork'.

Turnbull and Eresman then stopped off at the airport where all the planes, including 'the little de Havillands and everything were all scrunched up'.

Eresman dropped Ruth off at HMAS *Coonawarra* sometime before midday, and then proceeded to photograph more of the devastation that was Darwin. Ruth made her way to the old Wrans quarters where she had been billeted. Most of the top floor had been wrecked but she found that she 'still had a door'.

We Wrans and sailors gathered in the mess sometime during the afternoon of Christmas Day and registered our names with a warrant officer. All the phones were out of order and our radio station had been destroyed, but the naval authorities were attempting to inform our families that we were alive. When it was my turn to tell the very tired-looking WO what I wanted on the telegram to my mother, I could only think to say, 'Am okay.'

* * *

During the afternoon, *Arrow*'s CO Robert Dagworthy was finally reunited with his wife, Christine, and daughters, Siobhan and Kate, who were staying at the Larrakeyah Army Barracks. Camp stretchers were set up and Christine and her girls 'bedded down on those for a few nights'. She doesn't remember how many nights they stayed there. 'The days all blended in. We were brought essential supplies such as milk. I remember we had to improvise nappies for the children using damask serviettes from the officers mess lined with sanitary napkins. That was how we managed.'

* * *

Carolyn and Barry Spencer knew there wouldn't be much left of their house. Once they felt mentally able, they went out to the 11 Mile to check. 'All that remained of our house were the stilts,' recalls Carolyn. 'Everything had blown away, except the bedroom where the bed was still made and untouched. There wasn't much left to salvage. We gathered what we could into a few garbage bags and then had a long shower under a fire hydrant. Pure bliss!'

* * *

The patrol boat *Attack* was secured as firmly as possible with mooring lines around the rocks. That night, the ship's complement all climbed back up the cliffs and made their way to the sergeants mess at Larrakeyah. Part of the roof was missing off the building. Nevertheless, they were invited to come in and camp on the floor until further arrangements could be made for them.

CHAPTER 11

TAKING STOCK

Initially, our local naval forces could offer little assistance to the civilian population. Two of the four patrol boats were out of action, and most of the infrastructure of our naval bases, radio stations and headquarters had been destroyed. The oil fuel lighter had broken away from its moorings and went aground in the mud flats off Nightcliff.

All naval personnel were themselves victims of the cyclone, with many injuries and the loss of our homes and possessions. Worst by far were our casualties.

NOCNA, Captain Eric Johnston, reported that, 'POQMG L.K. Catton, R54753, and ABMTP I.R. Rennie, R107825, did not survive the sinking of HMAS *Arrow*; in addition, the wife and two children of ABETC G.W. Stephenson, R64881, and the youngest child of ABMTP S.W. Bruhn, R42996, were killed in their married quarters.'

On 26 May 1996, Lyn Alderton, my old roommate, reminded me in a letter that a memorial service was held outside the Admin Building at *Coonawarra* a day or so after the cyclone. We had sung the Naval Hymn, 'Eternal Father, Strong to Save'.

Lyn wrote, *I've been meaning to thank you for being there beside me – you were a tower of strength for me at that moment and I will never be able to thank you enough.* Her words surprised me because I'd been taking strength from her. She was a steadying influence on me then and remains so to this day. Since that memorial service, I have never been able to sing the Naval Hymn without breaking down into silent tears during the fifth line – 'Oh hear us when we cry to Thee, for those in peril on the sea'.

* * *

The two operational patrol boats, *Advance* and *Assail*, proceeded to conduct search and rescue operations in the harbour area, along Darwin's foreshores, and in the waters out to Melville Island. The wharves were cleared of storm debris and the water surveyed for wrecks.

* * *

Some Wrans had been off the base on Christmas Eve. They were unable to report for duty at HMAS *Coonawarra* the next day because so many roads were impassable due to debris. Instead, they assisted civilian relief organisations and manned communication centres for the police, fire brigade and Department of the Northern Territory. Other Wrans provided a twenty-four-hour courier service on an hourly basis between RAAF Darwin, *Coonawarra*, Shoal Bay and the civil authorities.

In his lecture 'Operation Navy Help', NOCNA Eric Johnston remarked:

> May I say in passing, and I hope this will not be taken as a chauvinistic comment, that one of the most delightful sights on Christmas Day and Boxing Day was my Wran couriers, who had lost the majority of their clothing, riding motor bikes through the city in the skimpiest of bikinis but none the less with their service

> caps planted firmly on their heads. This unusual garb was to earn me a severe censure from the fleet admiral on his subsequent arrival in Darwin, a censure which I took but lightly.

The local naval personnel spent Christmas Day clearing naval roads, estimating naval damage, searching the harbour foreshore for survivors and trying to establish communications with Canberra.

Because the Naval Headquarters building had been destroyed, NOCNA temporarily shifted his headquarters to Larrakeyah Army Barracks, where many naval families were sheltering. Some hours later, NOCNA closed down HMAS *Melville* and moved his headquarters to his residence, Admiralty House on the Esplanade. The naval personnel and dependants at HMAS *Melville* were moved to *Coonawarra*. Decisions were being made on the run as naval authorities adapted to the chaotic environment, establishing order in the aftermath of disaster.

Admiralty House became the centre for Darwin's post-cyclone reconstruction. Captain Johnston and his wife, Joan, moved into the flat beneath the house. The yard became a temporary Navy cookhouse, serving hundreds of meals a week.

Captain Johnston's kitchen clock was borrowed to mark Zulu time (Greenwich Mean Time) in a temporary communications centre.

By 8 pm on Christmas Day, good communications had been restored between NOCNA and Canberra. Captain Johnston issued a series of situation reports, specifying the damage to naval assets and giving an overview of the damage sustained by the city itself. Johnston also became a member of the Darwin Disaster Committee.

The top half of the junior sailors' accommodation at HMAS *Coonawarra* was destroyed, as was the upper floor of the old

Wrans quarters. The admin buildings were damaged and the houses in the marriage patch were damaged or demolished. Half of the transport building, which had been built during the Second World War, was demolished. The receiving station where I'd worked, also built during the Second World War, lost its roof and most of the external walls, and all the equipment was wrecked and water damaged.

Most, if not all, of the private vehicles of naval personnel were sandblasted, and many had broken windows or windscreens. The roads and footpaths were littered with debris, and the trees were bare of leaves.

The organisation started in earnest. Personnel were accounted for, hard rubbish was piled up, search and rescue operations were underway, and spaces were being made habitable for naval personnel and families. By late afternoon on Christmas Day, we were being officially assigned to clean-up and security duties. Many of us had been awake for thirty-six hours or so and some of us were sent off to the Junior Rates Club for a nap. I walked into the Rates and saw one sailor asleep propped up on a barstool and another asleep in a chair. That seemed a bit risky to me. I thought they could be injured if they stirred and fell. The floor seemed safer, so I lay down close to the wall and fell instantly asleep. I used to wonder how tired someone would have to be to sleep on the ground, without a pillow or rug. I now knew.

I was shaken awake after two hours to begin my watch on clean-up duties. We worked during the night, numbed with fatigue. While I was fortunate to have had a brief sleep, Wran Steward Joan Czarnecki was kept busy. She says:

> Bedding from the quarters was moved to a section of the main mess to provide accommodation for those who had lost their homes. There was clearing of cabins and facilities. I was tasked with the drying of blankets. Nearly everything which wasn't destroyed was wet.

> How long I spent in the NCOs' laundry loading the dryer, folding what came out and getting it moved to where it was needed, I can't remember. I finally got to put my head down on Boxing Day. My bed had been stripped by the wind and was part way across the room heading for the door. When I had left it, it had been lined up against the window. I used towels, one for a sheet to lie on, one for a blanket and one folded for a pillow; I was so tired. The last time I had slept was the night of the twenty-third and I wasn't the only one.

For a day or so, the watches were four hours on, four hours off. They were later reorganised to eight hours on, eight hours off.

* * *

On the afternoon of Christmas Day, the Emergency Services Committee held a meeting at police headquarters to discuss the response. The director of health, Dr Charlie Gurd, had worked through the night at the Darwin Hospital, and told the attendees, 'Look, we have no choice. We need to evacuate the population. It's got to be done.'

Dr Gurd had spent quite a bit of time in Africa and knew what consequences were likely to follow a disaster in a tropical location without hygiene, sewerage, water, electricity and with only limited medical facilities.

The other members of the committee agreed. He was right. Ansett Airline's NT manager recommended that the population be evacuated.

At 12:25 pm on Christmas Day, Alan O'Brien, the secretary of the Department of the Northern Territory, sent an official message through the Postmaster General (PMG) Communications Centre at Adelaide. The message read:

- Darwin completely devastated by cyclone last night. Deaths to date in neighbourhood of twenty. Ninety percent of houses completely or seriously damaged.
- Suggest national disaster team flown to Darwin immediately.
- Also suggest airlift of essential medical supplies, food, clothing, blankets, etc. commence immediately.
- Two surgical teams complete with supplies needed.
- Return flights may be used to solve probable evacuation problem.
- Director of Health has advised casualties may have to be evacuated.

As acting chairman of the Joint Service Local Planning Committee, NOCNA agreed with that assessment and recommended that all service dependants be evacuated as quickly as possible. There was very little accommodation at HMAS *Coonawarra* that was still usable and most of the families had nowhere to live.

A contributing factor was Darwin's remoteness. It was the most sizable city for a thousand or so kilometres, so supplies were not easily obtainable. Also, we were right in the middle of the Northern Territory's wet season, with the strong possibility of more storms and cyclones to come, and no safe shelters for the residents.

CHAPTER 12

IMMEDIATE RESPONSE

For the first few days after the cyclone, we Wrans and sailors in Darwin spent all our time doing physical labour or sleeping. There were no newspapers or TV and few of us had access to radios, so we were dependent on the grapevine for news. I, for one, was unaware at the time of the assistance being organised by the RAN.

When advised of the disaster, the Naval Air Station (NAS) *Nowra* in New South Wales swung into action.

The Navy operated two Hawker Siddeley HS 748 aircraft, which were medium-sized turboprop thirty-six-seat commercial airliners originally designed by the British firm Avro in the late 1950s. The Fleet Air Arm DC-3s began to be phased out in the early 1970s and were replaced by these two HS 748s. The first HS 748 was ferried by RAN personnel from Great Britain to Nowra in June 1973. The second aircraft arrived in August 1973, and both were allocated to 851 Squadron. They proved their worth in the years that followed. These two Navy aircraft were every bit as important to the RAN's relief efforts as the ships.

The first Navy HS 748 aircraft was flown by VC851 Squadron Commander, Lieutenant Commander James Campbell. It departed NAS *Nowra* at 8 am EST on Boxing Day for Sydney. Twenty Red Cross workers embarked with

200 units of blood and plasma. They then proceeded to Darwin, arriving at 4 pm CST.

The second aircraft, flown by Lieutenant Commander Peter Adams, departed NAS *Nowra* an hour after the first. It carried seven members of Clearance Diving Team One (CDT1) and Command Diving Officer, Lieutenant Commander T Jones, who had volunteered his services. Aircraft maintenance personnel with tools and spares were also on the flight. They arrived in Darwin two and a half hours after the first aircraft.

The flight approach to the Darwin airfield was made over a relatively unpopulated area, so the full devastation of the cyclone was not immediately apparent to the aircrews. However, the commanding officer of VC851 Squadron, Commander Jim Campbell, later said that the force of Tracy was evident when he saw 'a large white refrigerator firmly wedged in the branches of a tree about forty to fifty feet (twelve–fifteen metres) above the ground'.

There were no hangars at Darwin Airport for the aircraft maintenance personnel to work in, just an area of open concrete that radiated heat and humidity.

The two HS 748s of VC851 Squadron conducted a shuttle service until 10 January 1975. In the first ten days, they completed fourteen return flights, bringing specialist personnel and twenty-three tonnes of stores and equipment to Darwin. Among the cargo were pharmaceuticals, toys and magazines.

The return flights to Nowra were primarily filled with naval dependants from HMAS *Coonawarra* being evacuated south to family and friends, as well as naval personnel. The aircraft maintainers did double duty as flight stewards and converted the rear sections of the planes into an area for children to play and sleep during the long flight south. The evacuees were accommodated in HMAS *Kuttabul*, HMAS *Penguin* and HMAS *Watson* in Sydney, and at HMAS *Moreton* in Brisbane.

The maintenance personnel of HMAS *Albatross* then performed a twelve-hour overnight shift to service, prepare and load the planes with freight for an 8 am departure the next morning for Darwin.

In Darwin, time passed in waves of activity and fatigue. Michael French of HMAS *Coonawarra* recalls, 'The next few days [after the cyclone] were a bit of a blur with what I was doing. During the day, we were doing what we could with our own areas of responsibility. Planes were coming and we were helping to unload and distribute at night. I was getting back at night around 2200 hours (10 pm).'

Other naval personnel kept busy, and families coped as best they could while waiting to be evacuated. Liz Grimes recalls:

> On Boxing Day, a group of people were trying to find out the scores of the Boxing Day match. It came to me later on that we Darwinians, in the midst of disaster, were trying to find out the cricket scores on a tiny radio and could only pick up some station that wasn't giving out the score.
>
> I can't imagine the anguish my fellow Wrans went through at *Coonawarra*, being far from home and family. After the cyclone, Dad and I went to his property just south of Darwin to collect some fresh water. On the way, we stopped at *Coonawarra* to see how the Navy fared. I was very shocked to see the devastation to the Wrans block. I was assured that the girls were all okay but was saddened to hear two sailors off of HMAS *Arrow* had lost their lives.

Many people lost track of time. Christine Dagworthy, staying at the Larrakeyah Army Barracks, recalls:

> At some stage, possibly Boxing Day or the day after, I was taken back to the house in Nelson Street to try to

> salvage some of our belongings. We were still wearing the clothes that we were wearing on the night of the cyclone. Everything from the house had blown up the street. I vividly remember Kate's yellow cot was strewn in pieces. All that was left of the house was the floor and part of the bathroom.

There was more heartbreak to come. The Dagworthys lost their pet dog. Christine says:

> Germs, our dog, had been found earlier. She had been seriously injured and had been 'put down'. She had been a wonderful companion. We acquired Germs in Melbourne when Siobhan was a newborn baby. She always stayed close to Siobhan when she was in her playpen or pram in the garden in Melbourne. When we lived at South Hurstville in Sydney, we had a very large backyard. I would often be working up in the back of the garden while Siobhan and Kate were having their afternoon sleep. Germs would stay at the back door of the house. If either of the children woke up and had a little cry, she would run up to me and give a little 'woof' to attract my attention. She always alerted me when either of them woke up. Losing Germs was very sad, as she was devoted to Siobhan and Kate. She was a gentle, loving, protective friend.

The loss of cats and dogs, treasured companions, during the cyclone and in the weeks that followed, was hard to take. Many died as a result of the cyclone, and many were put down after being abandoned.

* * *

On arrival in Darwin on Boxing Day, the RAN's Clearance Diving Team One split into small groups and set to work at

Stokes Hill Wharf. Two of the patrol boats, *Assail* and *Advance*, were alongside the wharf. Both boats had propeller and shaft damage, but they were operational even though HMAS *Assail* only had one propeller.

The CDT1's urgent priority was to find HMAS *Arrow*. They also assisted in making *Attack* watertight and surveyed the harbour for wrecks. For weeks, they worked unstintingly in the harbour's murky black water filled with jagged metal, sunken boats and other debris. Michael Daley of HMAS *Advance* says:

> On Boxing Day, the clearance divers arrived in Darwin, and I had the task of feeding them out of my little galley, I think there was about fifteen or twenty in the end, but they ate like fifty after having worked all day. I was feeding our crew as well as the *Attack* crew for a few days. I was going like a cut cat for a couple of weeks there. The divers ate me out of house and home but supplied me with plenty of beer every night. The biggest issue I had was trying to get food. Our coxswain, Frank Warren, commandeered a vehicle and went out on the scrounge for whatever food he could beg, borrow or steal.

One night, Daly was having a few drinks at *Coonawarra* and wanted to return to *Advance*. The crew had been able to live on board their ship. Daly had no transport but spotted a police vehicle with the keys left in it. He happily drove that back to the wharf. The next day, the CO ordered him to 'Take it back!' He drove the police car back to *Coonawarra* and 'told the cops I found it down on the wharf. They were very grateful to me for returning it.'

Daly again had the problem of returning to his ship from *Coonawarra*. He found one of the old blue Navy Kombi vans parked by itself with the keys in it. He loaded up the van with food and drove back to the wharf. 'I used this vehicle for a few days and went around to *Coonawarra* and any produce storage sheds that were still standing to find any meat and vegetables that weren't spoiled, as they were giving it away,' Daly says. 'I had many mouths to feed, so in the early days after the cyclone, it was a constant hunt for food.'

Vehicles weren't too hard to get. Cars in car yards were all written off and many others were abandoned. Daly remembers, 'You could spray paint NAVY on the side, and no-one would question you.' Many of us used our initiative and salvaged whatever we could use in the clean-up effort.

Daly strived to feed the crews of *Advance* and *Attack* as well as the clearance divers, all of whom had big appetites from hard work. *Advance*'s CO, Lieutenant PB Breeze, reported that the ship's company spent weeks cleaning up the ship, searching Darwin Harbour for bodies and assisting search and clean-up teams in the city.

He noted that 'a hull survey carried out by CDT1 after the cyclone revealed no damage other than the starboard propeller already mentioned'. Repairs were achievable.

There's a saying that 'an army marches on its stomach' and that was relevant to us as well. Despite the heat and the stench, our appetites were whetted by hard physical work and the Navy cooks did a wonderful job of feeding us all under severely trying conditions. I don't remember ever going hungry.

* * *

A massive amount of work was needed on HMAS *Attack*, run aground on the rocks under the cliffs. Some of the naval personnel based at Larrakeyah were part of fleet maintenance, including a petty officer shipwright, who, according to Bernie Kloss:

> Came down at low tide when the *Attack* was high and dry, with wooden plugs, wedges, round plugs, all that

> kind of stuff, and fast-drying fibreglass. He banged the plugs into the different holes and tried to patch the hull up enough for us to get the vessel back in the water and around onto the slipway.

One of the biggest trawlers in Darwin, *Raima Kathleen*, was tasked with towing *Attack* off the rocks on Friday 27 December. A line was thrown across from the trawler to the patrol boat, and CO Lieutenant Paul de Graaff ordered Kloss and one of the ABs to stay on board and put their lifejackets on. Once on board the battered *Attack*, de Graaf said to Kloss, 'Right, you're out there to run around and check if there's any water coming in too fast.'

If that's the case, Kloss remembers thinking, *I'm not going down too far into the boat, I can tell you that!* A line was placed on the stern of *Raima Kathleen*, and once it was high tide, they gave 'a big tug', and 'there we were', says Kloss, 'drifting very low in the water because there was a lot of water on board'.

At 5 pm, *Raima Kathleen* dragged the patrol boat sideways until they were far enough out to manoeuvre the mooring line around so it was on the bow of the Navy vessel. The trawler then towed *Attack* to Stokes Hill Wharf, where it berthed at 6:30 pm. Everyone worked around the clock to pump out the water.

Earlier, the Navy clearance divers had worked unstintingly to locate HMAS *Arrow*, surveying Darwin Harbour. Kloss says:

> The divers went down underneath our boat, *Attack*, and patched the hull up with blankets and pillows and all that kind of stuff, and then they put straps right around the hull to hold it all in place.

Kloss estimates the divers worked for two to three days while he and other crew members tried to pump the water out of the damaged areas. He remembers, in particular, the forward store where the food was kept was a mess, with 'tea bags, noodles and everything else blocking up the suctions on our pump'. In an attempt to unblock them, they had to wade around 'in food, flour, and stuff. It was horrible.' More 'horrifying', says Kloss, was the 200 cartons under the petty officers mess that had gotten wet. 'The cardboard cartons all got soft and fell apart and floated away, and there's all these beer cans floating around.' Kloss and his crew retrieved them all and stacked them in the petty officers' shower and toilet for safekeeping.

Officially, *Attack* sustained 'two holes below the water line' after its grounding during Cyclone Tracy, as recorded in the report of proceedings. Lieutenant Paul de Graaf noted that one was 'four x three inches on the keel', and the other was also 'four x three inches on the starboard side'. He also wrote, 'Quite severe indentation of the starboard side has occurred and possible hairline fractures may exist. The cast iron base of the anchor winch had fractured around the securing bolts, the whole winch had broken free and was located near the main superstructure.'

Attack needed extensive repairs and would be out of action for many months to come.

* * *

Fortunately, the Naval Oil Fuel Installation at Stokes Hill suffered little structural damage. However, the capacity to supply fuel immediately after the cyclone was temporarily affected due to the ingress of water into electrical circuits in exposed positions. The old hose shed, which was due for demolition anyway, was destroyed. Bill Ritchie, an officer on HMAS *Melbourne*, comments:

> A lesson to come out of Darwin concerned the provision of fuel. Because the power grid was destroyed, it turned out there was only one outlet in the whole of Darwin that

> had the ability to pump fuel by hand. So much for the technical age. All the others had to wait until generators could be found or flown in from the south and connected to each petrol station.

Some of this information was trickling around our ranks. There didn't seem to be much good news. More and more problems were identified and prioritised in a seemingly never-ending list of matters needing urgent attention. The scale of the work needed was overwhelming. All I could do was focus on what was in front of me, on what I could manage to do myself. Like everybody else, I chipped away at the tasks assigned to me and gave a hand to my mates when needed. I frequently saw sailors and Wrans noticing what needed to be done and just getting on with it.

CHAPTER 13

THE CLEAN-UP

In the aftermath of Cyclone Tracy, sailors scoured the personnel quarters for as many plastic-covered mattresses as they could find and provided them to now homeless naval wives and children. As many others as possible were laid end to end in rows on the floor of the Rates Club. Sailors and Wrans slept on them in their brief times off. There weren't enough mattresses, even with only half the ships' company using them at any time, so everyone slept two to a single mattress. It made no difference to us whether we shared a mattress with another woman or man; we were all too tired to care.

One morning, a Wran gently shook me awake in the early hours and asked if I would mind moving to another mattress as I was sleeping with her husband, and she wanted to! I got up and, with bleary eyes, staggered up and down the rows before finding another half-vacant mattress.

Another night, I bunked down with a young sailor I'd never met before. We slept back-to-back and I remember thinking how very comfortable he was. A morning or so later in the junior sailors mess, I sat down for breakfast at a table with a bunch of sailors. In the general conversation, one of the sailors said to me, 'Don't I know you from somewhere?' and before I could respond he said, 'I know! We slept together the other night.'

The other sailors burst out laughing. They created an uproar, hooting and hollering and bashing their cutlery and cups on the table. The more the young sailor tried to explain, the more they drowned him out with catcalls and laughter. They all knew perfectly well what the situation had been but weren't going to let that stand in the way of a good stir. I nearly cried laughing and the young sailor and I both blushed crimson. I am still always amused when people use the euphemism 'sleeping with' someone when they mean having sex, because the reality for us in the days immediately after Tracy was that we were simply and literally sleeping with fellow Wrans and sailors.

* * *

With no electricity or running water in most of the buildings at *Coonawarra*, we were unable to wash and dry our clothes or footwear. Clean-up duties kept us too busy to attend to our own gear for a few days anyway.

The entire top floor of the old Wrans quarters had been levelled in the cyclone, and most of the Wrans who'd lived there were left with nothing. The second floor of the new Wrans quarters was uninhabitable as the roof was severely damaged. Most of the Wrans remaining at *Coonawarra* were relocated to the first floor of the new Wrans quarters. My room on the first floor had only sustained water damage but, in the high humidity, fabrics quickly grew mouldy, and my clothes and footwear became unsalvageable. We were clearing debris while wearing the remnants of our service uniforms and civilian clothes, with no boots, hats or protective gloves. Working hard in the oppressive heat and humidity meant our personal hygiene was almost impossible to maintain as well.

After a couple of days, a sailor turned on a fire hydrant in the lawn between the two Wrans quarters, and word went around fast. Some Wrans had managed to salvage their bikinis or swimming togs, but many of us just had to strip down to our bras and panties. We lined up and went through the water spray from the hydrant. Then we soaped up our bodies, clothes and hair with Gamophen soap. While waiting back in line to be rinsed off with the hydrant, we each scrubbed the back of the Wran in front. I had never felt so wonderfully clean. The sailor kept a straight face all the while, but everyone reckoned he must have had the best job in the Navy. To this day, the scent of Gamophen soap takes me straight back to the heady euphoria and relief of being fresh and clean at last.

* * *

Because HMAS *Coonawarra* was a communications station, most of the sailors and Wrans stationed there were radio operators. With our radio station out of commission and Shoal Bay still under construction, we were unable to perform our usual duties. Instead, we got stuck into whatever needed doing.

Some of us cleaned up the debris at *Coonawarra* and performed repairs and demolition duties. Wran Steward Joan Czarnecki says:

> To prevent the spread of disease, the stewards were tasked with going through abandoned marriage patch houses and emptying the freezers and fridges of rotting food. Note well, melted ice cream left in tropical heat for days smells worse than rotten fish!
>
> We'd knock on the doors and if no-one answered but the door opened, we went in and filled bins with stuff to be collected. If you walked into a yard and the dogs gave chase, you bolted and left that one to the occupants to sort out. I hadn't been very good at the hurdles in my school days, short legs, but at one such place there were two bull terriers (pig dogs), who had let me get almost to

the door. Well needless to say, I excelled that day and ran my personal best and cleanly hurdled the fence.

When most of the children had been evacuated south, there was a surplus of baby food. This was passed on to the Wrans for snacks when we couldn't get to the mess for proper meals. It really was a case of 'waste not, want not'.

'In my opinion, the fruit and jelly were good, the custard bland, but the meat and vegetable ones were blah!' says Czarnecki. 'Disgusting, no taste and no texture. When I had my own kids, I cooked and pureed my own baby food for them.'

Another task that Joan and other Wrans assisted with was spraying for insect pests such as flies and mosquitoes. There was no personal protective equipment (PPE) available to protect naval personnel from the chemicals used. They were clad only in shirts, shorts and sandals, with a lot of skin exposed.

Czarnecki says, 'Still to this day, I appreciate the sick bay attendant giving us surgical masks in some effort to protect us from inhaling the stuff. We went around standing on the trailer's frame and tongue, hanging on to the sides. We didn't get in with the drum and sprayers.'

Clearing debris was an imperative. It was still cyclone season so more storms were inevitable, and the prospect of more debris flying around was alarming. Wrans cleared rubble from the top floor of the old Wrans block. A dump truck pulled up close to the building on the road below, and Wrans set to with a will, hurling bricks, concrete, wood and tin into it. They put aside anything personal that could be salvaged, but there was little of value remaining. The diminutive Czarnecki remarks that someone gave her a sledgehammer, and she employed it with a vengeance, working off a lot of stress by bashing the remains of the brick walls into smaller pieces for ease of removal. She says, 'I had a really good workout with that, so satisfying.'

After a path had been cleared for traffic on the highway, Wran Radio Operator Dot Wright and I were assigned to a squad to requisition supplies from a warehouse in Darwin's industrial area. We travelled in a Navy bus capably driven by Chief Petty Officer Coxswain Rosenthal. The path cleared on the road was probably adequate for cars, but the bus frequently had difficulty negotiating around debris. Conditions were much worse when we reached the little-travelled industrial area.

The designated warehouse had only the rear wall standing, as well as part of the roof and two sidewalls. Most of the goods that hadn't been blown away were piled high in the middle. Dot, a few others and I climbed the pile, passing down any unspoiled goods that could be used. We collected mainly food supplies – tins of food and milk powder – but also came across fawn-coloured uniforms that had been destined for the Northern Territory Police. We were directed to collect them too.

When the sailors had packed the bus with supplies, we were ordered to get down from the pile and board the bus. Boxes were packed under the seats and up to the ceiling all along the sides of the bus, so we had to stand in the aisles on the journey back to the base. It was a hairy trip. We all stood with legs apart, bracing against the boxes on either side, trying to keep our balance and prevent the boxes from falling on us, while the CPO manoeuvred the bus around the debris and massive potholes. I thought it would be just my luck to survive the cyclone only to be killed by a carton of powdered milk.

The Northern Territory police shirts were distributed to naval officers and non-commissioned officers. Very few of them had serviceable uniforms, so the police shirts were an acceptable substitute. As lowly Wrans, Dot and I weren't entitled to a shirt, but we managed to score one each anyway. I still have mine, complete with a few small blood stains.

Sailors operated bulldozers and earth-moving equipment, clearing the grounds and roads on the base, piling the debris together. They worked unstintingly, soaked in sweat. Some of the most noted rabble-rousers, the best larrikins, were the most proficient workers. They worked tremendously hard, with no supervision necessary. They just saw what needed to be done and got on with it. So did the Wrans.

Some Wrans who had once been nurses helped in the sick bay. A petty officer and three Wrans operated the telephones at the Darwin Fire Station. Two who had been qualified telephonists in civilian life went to work at the telephone exchange.

One Wran radio operator who had been off the base on Christmas Eve was unable to return to *Coonawarra* the next day. She reported at the local police station and was immediately seconded to work there to help with communications. In an interview for the *NT News*, Major General Alan Stretton recalled, when he 'walked up the steps of the police station, it had been raining, [and] there was blood running down the corridors, corpses were laid out on the floor and women and children were sobbing in shock'. The Wran worked there in a room displaying the photographs of all the bodies that had been recovered – a harrowing experience which she handled with dignity and characteristic efficiency.

Another Wran was a transport officer for a week and others worked as dispatch riders. Two Wrans assisted at relief centres, and others toiled with sailors on clean-up duties in two of the worst-hit suburbs. Wran Radio Operator Ruth Turnbull was assigned to a crew clearing out houses on the marriage patch and around the Forbes Street area. Over forty years later, she still shuddered at the memory, telling me, 'We had to throw the carpets out, and they were putrid. And the maggots in the fridges! And the smell! Oh Jesus.'

* * *

Adding to the strain in the first few days after the cyclone was the fact that Les Catton was still missing, as was HMAS *Arrow*.

The water in the harbour was dark, murky and filled with debris. A team of clearance divers searched under the wharf, and other naval personnel scoured the mud flats and beaches. According to AB Cook Michael Daly of HMAS *Advance*, 'no-one knew exactly where *Arrow* lay', so one afternoon he and ABET Ray Minette decided to take out a ship's tinnie from Stokes Hill Wharf to see if they could find the sunken patrol boat.

'Ray drove the boat and I had a grappling hook,' Daly says. 'I was just throwing this thing in to see if I could catch something.'

After a while, Daly 'caught something which gave way' and so pulled on the hook:

> As I pulled it up, a body popped up face down. We rolled it over when tying a rope to its leg and we were shocked to discover that it was the buffer off *Arrow*, PO Les Catton. He only had his jocks on. He'd been in the water two or three days. I remember for weeks I couldn't get that slimy feeling of him out of my mind.

Daly used the radio they had on board to call up *Advance*, who then contacted the police. He and Minette steered the tinnie toward the entrance of the wharf, beaching there until the police arrived in a five-ton truck with a tray on the back. The two sailors had already pulled Catton out of the water and 'wrapped him in the tinnie's boat cover'. Daly recalls one of the policemen helped them load Catton's body onto the tray of the truck, after which they 'left the tinnie boat on the beach and went with him to make sure he was looked after'. The police drove them to a funeral parlour.

> I don't know what was going on in town but that's where he took us. Anyway, we carried Cats in and there were at least five bodies on the floor in the back of the

> funeral parlour. They had a row of fridges – there were probably only four or five fridges – and they had two bodies stacked one on top of each other in each fridge. That was a massive shock; that's a lot.

Daly and Minette helped put their mate on the deck and were reassured he'd be looked after. After getting a lift back to the tinnie, the two sailors returned to *Advance*. Daly says he 'washed [his] hands for about an hour to try and get the stink and slime off' and then had to go back to cooking. He'd met Catton once on a night out 'on the booze' and the experience of finding his body was 'pretty devastating'.

* * *

The looting started early on Christmas morning. Michael Daly recalls going into town where 'lots of shop owners would be sitting there or standing there with their rifles because they had issues with looters'. The Navy was also subject to looting. NOCNA Eric Johnston reported that it was 'necessary to set constant guards on such places as the wrecked Naval Headquarters'. On top of their already onerous duties, sailors from the patrol boats performed guard duties day and night at the Naval HQ. They were exhausted. Though it wasn't clear whether 'small items of value were destroyed by the cyclone or removed by looters', Johnston 'personally, on two successive nights, had to threaten looters with a shotgun to remove them from the grounds of Admiralty House'. He was confident, however, 'that no items of a classified nature fell into the hands of looters and in general they were kept clear of naval property'.

A few days after the cyclone, I was assigned duties at what was left of our old radio station at HMAS *Coonawarra*. All the confidential paperwork there couldn't be stored securely, so several Wran radio operators were tasked with destroying the paperwork by day and standing guard at night. It was sad to see that old building in ruins. The communications equipment that I and my fellow radio operators had worked on was knocked over, strewn around and covered by the collapsed ceiling and roof.

By day, two Wrans at a time worked together scrunching classified documents and feeding them into an outdoor incinerator. Coupled with the heat and humidity, standing in the sun and working with fire was uncomfortable. Our hands cramped a lot, but to me the simplicity of the task was curiously peaceful.

By night, we took turns on sentry duty, guarding the remaining classified records. There was no electricity at the station and the only illumination was provided by a lantern. Wrans in those days didn't do weapons drill, and I had never used a firearm before. On my first watch on sentry duty, I was given a rifle and shown the safety catch and other features. The watches were usually quiet, lonely and uneventful, except for when a security patrol came by on their rounds of the base. Somebody had given me a book, *Tales of the South Pacific* by James A Michener, and I read it every night by the dim light of the lantern. That helped me stay awake and was a welcome distraction from my thoughts.

I find the following difficult to write. I was unable to talk about it for about twenty years. The image is starkly clear, but it was hard to put into words.

One night, as I stood watch, a figure appeared in the darkness outside the wrecked station. I took up the rifle, walked outside, levelled the rifle at the intruder and challenged him with a 'Halt! Who goes there?' This was something I had picked up from movies, not having any formal training to fall back on.

The man froze, mid-stride. The whole image froze. He was just a dark figure about three metres away, sighted in my rifle. There was a roaring sound in my ears. We stood off. Then I bellowed something like 'Stand, or I fire!'

He turned and ran, disappearing into the darkness after only a few strides.

I stood in shuddering silence for some time. The roaring in my ears subsided and I strained to hear him returning. I eventually went back to my station and waited for the patrol to turn up, doing their rounds.

When they arrived I reported the incident to the patrol leader, Radio Supervisor Special 'Scoops' Cooper. He immediately sent the rest of the patrol out to search the area and the perimeter of the base. He and I chatted together until the patrol returned without finding anyone. RSS Cooper said the man had to be a looter. If he had been a serviceman, he would have identified himself immediately.

The patrol left to continue their rounds. I was left alone with my thoughts.

I was nineteen, the product of an all-girls school. I'd just survived a cyclone, was exhausted, had no weapons training, was in the dark in the middle of the night, on my own, with no communications and no help within shouting distance.

Thoughts kept rolling around in my mind. What if he hadn't run off? What if he hadn't been alone? Should I have fired a warning shot over his head? What about taking the safety catch off? If I'd fumbled with it, he would have known that I was untrained with firearms. He might have then had the confidence to attack. If I had had to fire at him, would it have been best to try to shoot him in the leg or the chest? Assuming my aim was in any way accurate, which wasn't likely. Could I live with having shot somebody? Hang on, was the rifle even loaded? To this day, I can't remember if it was or not. Probably not. And in the dark, the thoughts started turning around all over again.

For years, I was haunted by those overwhelming feelings of fear and inadequacy. In my early thirties, I became an ambulance officer. Many of the male ambos at my station frequently told me that women had no place in the service.

According to them, women weren't up to the job. We weren't strong enough and would only be a hindrance in violent situations. For years, I didn't finish a single shift without being put down by at least one male officer or male patient.

I thought I was a capable person, handled the work well and really loved patient care, but whenever I fired up and defended myself against the men's remarks, I would remember the fear and inadequacy I felt during that confrontation with the looter and would doubt myself. I came to push myself very hard over the years, to overcome those feelings, that image. It took a long time for me to fully appreciate the importance of the actions I took at the time, rather than how I felt, and to find respect within myself.

The fact was, I had been assigned the duty of safeguarding classified material at the wrecked radio station. I challenged an intruder, scared him off and secured the station. I was desperately tired, frightened, untrained and unprepared, but I did my duty. It took me twenty-three years to look at the incident from that perspective.

Perhaps early counselling would have given me the more appropriate perspective much sooner – however, I'm grateful for the belated insight. I'm also grateful for the confidence the Navy had in the Wrans. We were expected to do our duty, and that's exactly what we did.

CHAPTER 14

EVACUATION

Just prior to the cyclone, Darwin's population was estimated to be about 47,000. With essential services severed, minimum food and shelter, and the risk of disease, Darwin could barely support a fraction of that number. Consequently, around 42,000 people were evacuated.

Many people left of their own accord by road, and others were evacuated compulsorily by aircraft. The airlift began on Boxing Day and over the next six days more than 25,000 were evacuated to the southern cities. Darwin's population was eventually reduced to a little more than 10,000, including those who came up to assist. For the next six months, access to the city was regulated by means of a permit system.

Wrans based in Darwin were given the choice to evacuate or stay. About thirty to forty Wrans, including me, stayed and worked on clean-up duties alongside the sailors. We were scattered and had a variety of tasks around the town, so I don't know how many of us there actually were.

The Army and RAAF evacuated their servicewomen as well as the wives and children of their servicemen. No exceptions. I remember a Wran who was married to a RAAF serviceman in Darwin turning up at *Coonawarra* in an absolute fury a few days after Tracy. She told us that the RAAF had insisted on evacuating her despite her strong protests that she was a serving member of the WRANS and wanted to report for duty at *Coonawarra*. She told us that as soon as her plane landed in Sydney, she caught the next flight straight back to Darwin. She was outraged that the RAAF only regarded her as the wife of a RAAF serviceman and would not respect her own status as a Wran. She was determined to do her duty alongside the rest of us.

* * *

On 20 May 1975, Rear Admiral Neil McDonald issued a report on 'the assistance and aid given by the Command of Flag Officer Commanding, East Australia Area, to the people of Darwin, the fleet and other organisations'. In his summary of events pertaining to the evacuation, he noted that on Thursday 26 December 1974, Boxing Day, 'All East Australia Area establishments were instructed to report availability of accommodation for Darwin evacuees.'

Personnel at the naval establishment HMAS *Kuttabul* in Sydney and members of the Navy Wives Association promptly prepared beds, toilets, food, telephone and telegraph facilities.

Pat Nuss was among the many Wrans who rose to the challenge of dealing with the influx of evacuees. Pat was a petty officer serving in the family services section of the Command Personal Services Organisation (CPSO) in Sydney. She was having dinner on Christmas Day with her husband Paul, also a petty officer, and her mother when they heard the news that a cyclone had hit Darwin and all contact had been lost. Paul immediately put up the ironing board and started ironing his rigs in preparation for duty. Pat rang the duty officer at Garden Island and gave him her contact details. Pat was then told that 'if my husband was in the Navy for him to report ASAP – to which I replied he was just packing and would be there within the hour'.

Pat drove Paul to Garden Island to join his ship, HMAS *Brisbane* (D41).

Later, Pat worked with the Navy Wives Association, preparing accommodation at HMAS *Kuttabul* for naval families being evacuated from Darwin. She says:

> I think it is important to mention this now-defunct organisation, as there were admiral wives and commodore wives with me making beds etc., and they were always supportive of CPSO functions for families of deployed members. The bedding store threw open its doors and handed out bedding and equipment without any fuss – I don't think their stocktake list balanced after Tracy!

At about thirty minutes past midnight on Friday 27 December 1974, the first group of evacuees airlifted from Darwin arrived at Mascot Airport in Sydney. They were reportedly met by a reception of Wrans, social workers and medical staff. A team had been organised to take care of baggage, but evacuees had very few belongings.

Lieutenant Commander (LCDR) WOC Roberts and LCDR MH Dowsett found it difficult at first to establish a base of operations at Mascot. Planning and liaison were hampered by the initial lack of coordination of the many organisations that had representatives at the airport, and information about when and where aircraft carrying evacuees would land was scant. For reasons that completely escape me, the Department of Housing and Construction became the designated coordinating authority for evacuees at the airport.

After the arrival of the first group of evacuees, RAN reception teams served six- to eight-hour shifts at Mascot. At the peak of the airlift, the teams comprised one officer, one senior sailor, a Wrans officer, six Wrans, a first-aid medical team of three, six buses and drivers, and two radio cars plus drivers.

The mammoth airlift of evacuees from Darwin involved Naval, RAAF, USAF, RNZAF and civilian aircraft. During the night of 28–29 December, a reported thirteen aircraft landed at Mascot Airport in a six-hour period, carrying an estimated 1300 evacuees. Their greatest challenge was when a QANTAS Boeing 747 arrived with 642 passengers on board.

The command liaison teams met all the arrivals. It was difficult for these reception teams to identify naval personnel and dependants among the incoming passengers, but fortunately the teams' Navy uniforms served as a beacon for the naval evacuees. A medical party checked arrivals, and transport was provided to the naval establishments HMAS *Kuttabul*, HMAS *Penguin* and HMAS *Watson*. The naval bases HMAS *Nirimba* and HMAS *Albatross* would only be used if the nearer establishments could not accommodate all the evacuees allocated to the Navy. Pat Nuss elaborates:

> I did shifts at the airport to meet incoming flights, and as they disembarked, families were welcomed, details taken, light refreshments offered and [they were] assisted around tables of clothing. Some families had some clothes with them, but it was obvious that they were in shock as they packed, as I remember what they brought with them was odd – one shoe of a pair, pieces of husband's clothes and other odd bits and pieces. They were able to collect enough emergency clothing for them and their children for a few days.

The flag officer reported that, 'The naval social workers and the Command Personal Services Organisation assisted in both the airport and the reception centres established in HMAS *Kuttabul*, *Watson* and *Penguin*.' The evacuees were all in a state of shock and desperately tired. They didn't know what to do next or where to go. They had little or no money, no home and few belongings. Some were still in the nighties they'd been wearing on Christmas Eve. Mothers had the extra pressure of caring for their children and trying to keep them calm and entertained, while their own world had fallen apart. The reception teams provided immediate security and reassurance.

The evacuees were processed, and most were then transported to HMAS *Kuttabul*, where they were provided with hot drinks, showers and bedrooms. It quickly became clear that more staff would be required to facilitate transit accommodation and further transport for evacuees, so at 4 am on Friday 27 December, it was decided to recall all members of *Kuttabul*'s ship's company. The recall was only broadcast in Sydney, so many personnel who were interstate didn't hear that they were needed. Many other members of *Kuttabul*'s ship's company had volunteered to fill vacancies in the ships being prepared to sail to Darwin. All available personnel were assisted by volunteers and the reception organisation functioned well.

A whole floor of four-berth cabins at *Kuttabul* had been prepared for the women and children evacuated from Darwin. *Kuttabul* had a plentiful supply of blankets and pillows, but a call went out for additional sheets, pillowcases and towels.

The evacuees appeared at first sight to be unharmed, but medical assistance such as first aid proved necessary for many. One young mother, eight months pregnant with her second child, needed assessment and medical attention. Unfortunately, the children who were used to drinking Darwin water became ill with gastro-intestinal problems after drinking reticulated Sydney water. From then on, boiled water was provided.

Few evacuees had any belongings other than the clothes they were wearing. Where possible, those clothes were washed and dried for them by Wrans and volunteer wives. Clean clothes, toiletries, cots, baby food, toys and various other necessities were provided.

Social workers assisted the women and children. They interviewed adult evacuees, compiled as much useful information as they could, and gave them the opportunity to contact family and friends. Wrans and women of the now-defunct Navy Wives Association were unstinting in the aid they provided to the evacuees, most of whom only stayed for a few hours.

By Friday evening, only twenty-one remained to be accommodated overnight. Naval personnel prepared to receive an influx of evacuees but no more stayed at *Kuttabul* until late on 30 December when three adults and two children were accommodated overnight. Pat Nuss shuttled back and forth between the airport and HMAS *Kuttabul*, which she described as 'all part of the job' to assess 'where families needed to go for support and assisting with onward travel'. When families were ready, they were put in taxis and flown to their next of kin. Nuss adds, 'I remember feeling very proud of the Navy, civilian members and the Navy Wives Association of Sydney for the way they threw themselves into any job that needed to be done. The whole exercise was such a success.'

The liaison teams at Mascot Airport were withdrawn on Monday 30 December as the number of evacuees being placed in naval establishments had reduced to a trickle. It is hard to imagine how desolated and lost evacuees would have felt on landing at the airport in Sydney without the immediate welcome and travel arrangements made by the reception teams, and then the further practical help provided at *Kuttabul*. I and many others consider these naval personnel and social workers an integral part of what became Operation Navy Help Darwin.

* * *

Many naval personnel and family members shared with me their memories of their evacuations from Darwin. They went by RAAF Hercules aircraft, Navy HS 748 planes, civilian airlines and by road.

Christine Dagworthy and her two daughters, two-year-old Siobhan and eight-month-old Kate, stayed at Larrakeyah Army Base for a few days after the cyclone. Their families were unaware of the devastation Tracy had wreaked on Darwin until the daughter of a family friend rang Christine's parents on Christmas Day in the afternoon. The friend had heard an ABC news 'report that *Arrow* had blown up' and 'lives had been lost on the ship'. Christine says:

> My parents didn't sleep that night. They were absolutely devastated. They presumed that Bob had lost his life. They feared the worst. Dad phoned his local federal member of parliament, Bill Morrison, who he knew personally, to ask if he had any details of what had happened to Bob and to us, but of course at that stage there was no communication. For over 24 hours, they thought that Bob had been killed, but had no news of what happened to us.

Her parents finally received a phone call 'in the early hours of 27 December' informing them Bob was alive, and Christine and the children were safe and being evacuated to Sydney. Christine and her daughters left Darwin in a HS 748, a naval turboprop aircraft, with other wives and children of naval personnel from *Coonawarra* and the patrol boats. 'Siobhan still remembers travelling on the aircraft with one of the Navy wives as I was nursing Kate,' says Christine. 'It's still an unhappy memory for her, of being separated from Kate and me.'

When the plane stopped over in Mount Isa, the evacuees entered the airport to find tables of clothes laid out for them. Christine picked out some singlets and shorts for her girls. She then accessed a public phone and rang her parents, apologising for reversing the charges as she had no money. She reassured her parents that they 'were uninjured, and Bob was safe' and 'would be able to come back to Sydney at some stage'.

From Mount Isa, Christine and the other evacuees were flown to the old international terminal in Sydney. By that stage, a Qantas Boeing 747 evacuating families from Darwin had set a world record for the number of people on board. There was some confusion about the size of the Navy HS 748 as some civilian authorities surmised that 'a 748 was one bigger than a 747', so a fleet of ambulances, police cars and a double decker bus were sent out to meet the HS 748 – a 36-seater aircraft.

Christine's father met her, Siobhan and Kate at the airport. Christine recalls:

> Coming back to Sydney, I felt safe and secure. My sisters Margaret and Elizabeth and their families had gathered at my parents' home at Bexley North to welcome us home. They said we looked like waifs when we arrived. My sisters put Siobhan and Kate in the bath together and remember removing tiny pieces of glass from Kate's hair.

They were also able to visit Christine's mother-in-law, who lived in Coogee, and reassure her they were uninjured and that 'Bob was well' and would return to Sydney 'when it was possible'.

* * *

Petty Officer Mick Hamilton had sustained serious injuries when the old transmission station collapsed around him. He was evacuated in a RAAF Hercules, a C130, from Darwin to the surgical ward in the hospital at HMAS *Penguin* in Sydney. He has little memory of the details of that trip, apart from being in extreme pain, barely conscious and on a stretcher. Sometime later, he became aware of a TV set in front of his hospital bed and watched 'Thommo and Lillee throwing them down against the Poms' in the Ashes Test being played in Melbourne.

Hamilton's wife, Doreen, had also been evacuated and was staying at her parents' place in Blacktown. When Doreen came to visit him, Hamilton asked her to smuggle in half a dozen beers, some Jatz biscuits and dip for him and the other patients to enjoy while they watched the cricket.

'And she did!' he says. 'I got sprung with the beer and I got run in. I had a bedside kangaroo court and summary trial, and my punishment was to clean the heads and showers each day.' As he couldn't get out of bed, he 'never actually fulfilled that punishment'. When he was able to get out and about, he escaped to the senior sailors mess and got caught and run in again. You really can't keep a good man down.

* * *

The wet season continued to make conditions uncomfortable for those of us remaining in Darwin. Civilian technician Michael Pilkington, his wife, Christine, and their two young sons were surrounded by debris at their home in Casuarina. Christine says, 'There were more storms afterward. We put up our tent the night after Tracy and I can remember lying in bed in the tent and physically shaking with fear. It was scary. Really, really scary.'

They wrote, 'Pilkingtons all ok' on what was left of the wall at the front of their house, so friends would know they were safe. The distress evident in her voice, Christine recalls:

> But one of the worst things was when the police came around and dragged all the dogs and took them down to Wagaman [a suburb of Darwin]. They were shipping people out, evacuating them, and people had to leave their dogs behind. The police were shooting the dogs and

> you could hear it. Alex next-door had a dog called Ears – a big black thing with big ears – and he wouldn't let the police take his dog.

Alex told the police that he refused to be evacuated and was going to stay in Darwin and care for his dog himself. And he did stay. The sounds of the dogs being shot stayed with Christine. She says, 'And that was awful, hearing that.'

The Navy told Michael and Christine Pilkington that their house was uninhabitable and moved Christine and the two boys to the undamaged house of friends at HMAS *Coonawarra*. Christine remembers going into the base and 'getting a washing machine to use'. She was then told she and the boys were to be evacuated without Michael.

'I was screaming my head off, and Michael said, "You've got to go".'

Christine replied, 'I'm not going. I'm staying here.' She didn't want to leave 'because I'm pretty tough', she says, 'but they were worried about disease'.

Michael told her, 'You've got to look after the boys.' With that, she left. The authorities flew them down 'in one of those bloody Hercules things. What an experience. It was shaking all the way down there.'

'Down there' was South Australia, where they stayed with Christine's mum at Henley Beach for three months. Meanwhile, Michael went to work at the Shoal Bay Communications Station, living in makeshift accommodation out there. In the evening, he would visit his next-door neighbour, taking him beers, food and other provisions.

* * *

On Christmas Day, Warrant Officer Keith Chay and his wife, Kathy, had driven their panel van from Nightcliff to HMAS *Coonawarra*. A few hours after arriving, Chay noticed his 'leg had become seriously inflamed'. He went to the sick bay, where they informed him his leg 'was badly infected from a cut [he] sustained from a glass louvre' while he was crawling to a neighbour's shelter. There was no penicillin available. As Chay was due to be transferred in a few weeks anyway, the XO suggested he travel to Katherine Hospital by car 'for an injection of penicillin', then continue on to his posting in Sydney.

Lacerations were notoriously difficult to heal in the hot, humid weather of Darwin in the wet season. Wounds often became ulcerated, and it wasn't unusual for servicemen and servicewomen to be sent south for treatment. The dangers of infection were greatly increased after Tracy and stocks of antibiotics were quickly exhausted. Fortunately, Chay's injury cleared up within a few days of his injection in Katherine, and his and Kathy's trip south was uneventful.

'It was wonderful to get back to our parents and families who had been extremely worried about our safety,' Keith says. He remembers with deep gratitude the organisations such as the Salvation Army, and the many Australians who offered aid and who 'donated both financially and with clothing and furniture'.

* * *

Petty Officer Liz Grimes was in Darwin on leave when Cyclone Tracy hit and was required to return to HMAS *Cerberus* in Victoria, where she was stationed. After her mother, sister and two nieces were evacuated to Brisbane, Liz reported to HMAS *Coonawarra*, where she spent a few days in the senior Wrans quarters before being evacuated back to *Cerberus*.

> I was flown out in a Navy plane, a HS 748. The trip was long as we landed at Tindal RAAF Base, Mt Isa and then the long haul to Sydney. At Sydney, I was transferred to a domestic flight, the plane was even delayed for me

to board. I was met at Melbourne by the Navy and taken to Greenwich House for an overnight stay. The next morning, I and another Wran returning from leave were transported to HMAS *Cerberus*.

* * *

Carolyn Spencer resumed her role as a civilian admin officer at HMAS *Coonawarra* and assisted with organising dependants to fly out from Darwin as soon as possible. Her mother, who had insisted on washing and helping while they were staying at CPO Rosenthal's house, was also on her way home.

Along with her other duties, Spencer organised first-aid kits for the *Arrow* crew who she would join when they drove in convoy to Alice Springs before flying to their homes.

'Still coming to grips with what had happened,' she says, 'I went to the sick bay and asked them to prepare first-aid kits, but to ensure wet checks – at least half a dozen packets – be put in.' She got 'a funny look from the Wran' before realising she meant to say Wet Ones.

After his wife, Christine, and two daughters were evacuated, Robert Dagworthy, CO of *Arrow*, stayed on in Darwin with the crew for about a week, assisting with the clean-up. He then joined that convoy of naval personnel and their families who drove in their own cars to Alice Springs. Carolyn Spencer says, after leaving Darwin, the convoy of *Arrow*'s crew were 'met all the way by Rotary, being fed and watered, given suitcases, fresh clothes and generally being well looked after'. At one township, after 'a lovely morning tea' organised by the CWA, Spencer realised she'd left her bag in the last town. It was flown to the next township for her to pick up there.

The bitumen road ended around a hundred kilometres south of Alice Springs, and the dirt road from there to Adelaide was heavily corrugated and damaged by road-train traffic. Most people drove from Darwin to the Alice, loaded their cars onto the train and either travelled in relative comfort on The Ghan to Adelaide or departed by plane. When the convoy arrived at Alice Springs, Dagworthy says, 'the cars were put on a train, and we were flown back to Sydney or other home ports'.

Rotary organised to have a special cage made for the flight that was big enough to accommodate Mandy, Spencer's Old English Sheep Dog. She ended up beating Spencer back to Sydney.

* * *

Other evacuations didn't always go according to plan for naval personnel and their families. Petty Officer Michael French says the Navy would only evacuate his wife and children by plane, but not his parents or the family dog. His father had managed to salvage French's car and caravan from beneath the debris at his house, so French's parents and his wife, Rosemary, decided to drive out of Darwin together, along with another family. French had to wait four months before he got leave to go to Melbourne and see his family again.

Rosemary captured their experiences in a poem.

MEMORIES OF CYCLONE TRACY

Rosemary French

T'was a long time ago,
That we first heard the blow
of that Cat 4 Tracy in Darwin.
We had family up for Christmas hols
so extra provisions were sourced
which turned out to be a waste
as we had to leave it all in haste.
You've probably heard that the sound of the wind was like
a locomotive,
it was indeed very loud but then we were very emotive.

Morning dawned and the damage was complete, not a
leaf on a tree was seen,
everything that was brown was now completely green …
less.
Deeds of bravery told to young and old with a nip o' rum
for breakfast.
The long and the short, if I had to report, that Tracy was
an experience to be lived.
Sadly, some did not live through to tell the tale, which
saddens a lot of folks.
As our home was trashed, we evacuated via 'The Track'
with mother-in-law, father-in-law, two sons, a dog and
myself.
More brave and heroic acts along the way south
With caravan in tow and possessions low,
I had my first gin and tonic in the Alice,
Was the best thing I've ever tasted as it hit the 'spot',
Mmmmm.
We flew from there to Melbourne, back to my parents'
place.
My mum and dad were very glad.
Funny thing, just because you know you are alright, loved
ones don't,
No mobile phones in those days, so they were extremely
relieved when we stepped in the door, to finally give
us a hug.

PART 3

OPERATION NAVY HELP DARWIN

'Since you all know your duties, gentlemen, I think you should start your preparations. The sun is about to set and you will find you have plenty to do.'

CS Forester, *Mr Midshipman Hornblower*

CHAPTER 15

ACTIVATING THE FLEET

Wednesday 25 December 1974 – Christmas Day

Early on Christmas Day, before communications were lost, a Darwin police officer phoned Major General Stretton, director of the recently established National Disasters Organisation (NDO) in Canberra. Over the noise of the storm, the police officer shouted, 'Darwin's been blown to pieces … for God's sake, send help!'

Further communications were lost for the next few hours, but according to legend, that was enough for the NDO to instigate a national emergency relief operation. The office of the Flag Officer Commanding East Australia Area (FOCEA) was notified at 7:30 am Eastern Standard Time (EST) that Darwin had no communications on any circuit. Shortly after 1:30 pm EST, reports of the extent of damage in Darwin began to filter through via the media, but there was no contact whatsoever from naval ships or establishments in Darwin.

Roger Fairbairn, Commander RAN (Retired), was a lieutenant on the staff of FOCEA in 1974. As the most junior officer, he was the designated duty officer on Christmas Eve and Christmas Day. He'd been posted back to FOCEA Naval Headquarters at the end of 1973 as the secretary to chief of staff (COS), Commodore AN Dollard. At the time, Rear Admiral WJ Dovers was FOCEA. Fairbairn's job was to:

> coordinate the matters and activities of the chief of staff that included all the land assets, all the Naval bases, the bombardment and missile range at Beacroft Head Jervis Bay, Garden Island dockyard, RAN store facilities at Piermont and Randwick where all ship spare parts were held, the ammunition depot at Newington, and other places such as the torpedo factory Neutral Bay and the range at Pittwater.

He spent most of his 'time organising meetings and co-ordinating with responsible folks on how things were working, solving, and escalating problems' and was 'responsible for coordinating the administration for the master plan for Garden Island'.

With all the senior officers having gone on leave the week before Christmas, Fairbairn was to be at Naval Headquarters, Sydney, until late January 1975 when he had leave scheduled. That was cancelled as a result of the response required by the Darwin disaster.

On Christmas morning, 1974, Fairbairn was in the officers quarters at HMAS *Kuttabul*, over the road from Naval Headquarters. He was called by the duty steward to the open telephone line mid-morning to answer a call from NOCNA Captain Eric Johnston, who he knew well.

NOCNA said, 'It's Captain Johnston, who's speaking?'

When Fairbairn identified himself, Johnston said, 'Oh you poor bastard.'

Johnston had been based temporarily at Naval Headquarters at one stage and used to sit in Fairbairn's office to chat and tell jokes. Lively sense of humour aside, Fairbairn believes Johnston 'was the right man at the right place'. Fairbairn says Johnston 'had just dragged himself out of the rubble of Naval Headquarters that had been destroyed by the powerful Cyclone Tracy'. Johnston explained to Fairbairn that 'he was talking on a temporary radio link to a communications facility in the centre of Australia', which was 'hooked to a telephone line' to Naval Headquarters in Sydney.

'It was the only number he could remember as everything had been destroyed in his office,' adds Fairbairn. Using the only method of communication from Darwin, Johnston gave Fairbairn a general description of events. Fairbairn scribbled down a few notes but committed most of it to memory. He had difficulty getting his head around such an unprecedented situation. He recalls Johnston saying to him:

> 'Keep in contact, do what you can, and we'll keep feeding you what the most important things are … I know you know everything that goes on in that place, good and bad. Just get it done.'
>
> Fairbairn replied, 'No problems. I've got the power to do that, and I don't care what the rules are; I will just get it done.'

As soon as Captain Johnston disconnected, Fairbairn tried to contact COS Commodore Dollard; however, Dollard was on leave and not at home. He also tried to contact FOCEA, Admiral Dovers, but he was also away. Those were the days before mobile phones, pager systems or emails, and people could only be contacted by a call on their home phone or a knock on the door.

Junior Lieutenant Fairbairn was left with the responsibility of coordinating Naval Headquarters Sydney's initial response to a national disaster. I can only imagine how aghast he must have felt, facing such an undertaking alone. He decided to call Lieutenant Commander WOC Roberts (known as 'WOC') of the RAN Reserve. Roberts was on leave at his cattle station in New South Wales and immediately said he would drive back to Sydney. Fairbairn said that Roberts was a distinguished Korean War veteran who had sound common sense. He'd worked with Roberts before and knew he would be 'the best guy to have' in the crisis.

Fairbairn called the duty and operations officers at Fleet Headquarters and told them about Johnston's call. The three duty officers exchanged views on what had to be done and agreed, 'Well, you do your thing and I'll do mine!'

Lieutenant Fairbairn went over to Naval Headquarters, opened it up and sat at his boss's desk, trying to think of exactly what to do next. He says:

> There were no Standard Operating Procedures for any sort of similar situation to manage. There was a security safe with some procedures and there were additional items in the HMAS *Kuttabul* secure communications centre that I also had access to; however, I knew that there were no guidelines to help and no other officers to assist me.

The best thing to do, he decided, would be to call the duty officer at Navy Office Canberra and advise him of the situation in Darwin as he knew it. That duty officer said he would inform senior officers in Canberra. Fairbairn then rang all the duty officers in all the various locations around Sydney – *Watson*, *Rushcutter*, *Waterhen*, *Penguin* and *Platypus* – and told them what had happened, advising them to recall personnel as quickly as possible and to stand by for further information. All those phone calls took several hours. He also rang a good friend, Ron Aitken, who was in the C-130 Hercules Squadron at RAAF Richmond and gave him the news to pass along to his CO.

Fairbairn started to 'send and receive signals using a comms sailor who ran up and down the street for about 150 metres between Naval Headquarters and the *Kuttabul* comms centre. The turnaround time for any communication was at least thirty minutes!'

He needed to record all the decisions made and actions taken. Surplus items were stored in the cellars of Naval

Headquarters, and when he'd been pottering around down there a week or so before, he'd seen some blackboards. He and two naval policemen relocated the blackboards to the chief of staff's office, which was quite large with bay windows. The office had a panoramic view stretching from the Harbour Bridge, Browns Wharf, the entrance to Garden Island to the harbour entrance to the Dry Dock.

Fairbairn set up the blackboards but had no chalk. He rang the wardroom and asked for one of the stewards to buy some chalk from the newsagent in Kings Cross. Amazingly, the shop was open on the public holiday. With blackboards and chalk, Fairbairn started recording the actions taken in response to the disaster.

Fairbairn's normal FOCEA staff officer duties were to know 'what was where and how to get it; where the boats were to do it; where was a crane; where were trucks that could do this, and who could say yes and how'. He knew 'where all the transportation resources were, including work boats and lighters at Garden Island that were used to ferry items around Sydney Harbour'. That knowledge was vital at the time because the organisation and coordination of the response to the disaster was totally ad hoc.

* * *

As the hours went by, staff at Fleet Headquarters decided to recall the crew of all the ships alongside Garden Island. After discussion with the duty officers at the FOCEA bases, they decided to recall all base personnel immediately as well. Radio and television stations around Australia broadcast the orders for all Navy, Army and Air Force personnel to immediately return to their bases and ships.

In Sydney, the usual Christmas routine had most of the fleet in harbour and undergoing self-maintenance. About half of the ships' personnel were away on annual leave throughout

Australia, and another third were ashore on short leave. Apart from the duty destroyer, the ships only had enough crews to maintain an efficient fire and damage control organisation. Machinery was stripped down for maintenance.

The Royal Australian Navy issued a general recall of all personnel to man HMA Ships *Melbourne*, *Brisbane*, *Stuart*, *Stalwart*, *Supply*, *Vendetta* and *Hobart* to sail to Darwin. The aircrews of 725 and 817 Squadrons were also recalled. Naval personnel on Christmas leave around the country responded immediately and left their homes and families to return to their ships.

William 'Bill' Ritchie was a lieutenant commander on the aircraft carrier HMAS *Melbourne* and was on duty on Christmas Day. 'For my sins,' he says, 'I had accepted responsibility as the duty officer on HMAS *Melbourne*' as the rostered officer was on compassionate leave. Ritchie's family had come aboard *Melbourne* for Christmas lunch, leaving the ship at about 2:30 pm. Ritchie returned to his cabin and shortly after, the steward knocked on his door to tell him the ship's Commander Air Group was on the phone in the wardroom. Answering the phone, Ritchie had expected a pleasant greeting but the commander got right to it and asked what was happening on the ship.

Ritchie said, 'Everything's quiet, and I've just seen my family off the ship.' He then asked what was going on.

The commander replied, 'The helicopter squadron has been recalled.'

When Ritchie asked why, he was told, 'They're going to Darwin. Didn't you know that Darwin's been wiped out?' and the commander followed up with details of the destruction.

HMAS *Melbourne* was the only vessel that could take the helicopter squadron to Darwin. Ritchie's mind swung into gear, and he considered what was necessary to prepare the ship for sailing.

The ship was undergoing maintenance and repairs and was in no way ready to go to sea. Where to start? Ritchie recalls:

> As I left the wardroom, the duty engineering officer was passing in the corridor and I asked him, 'What was the state of the engines?'
>
> His reply, 'In bits.'
>
> Me: 'How long to put them together again?'
>
> He: 'Three weeks with dockyard assistance.'
>
> Me: 'What about 0800 tomorrow.'
>
> He: 'No way in hell.'

Ritchie rang Fleet Headquarters, identified himself, and asked what the situation was concerning the ship and the helicopter recall. The reply was, 'We are too busy to talk to you.' Just as he hung up the phone, *Melbourne*'s captain rang enquiring what he knew. Ritchie told him what he had done so far, and that Fleet HQ was being most unhelpful. The captain said he would get back to him. Ritchie says:

> As you can appreciate, I was aware that all hell would be breaking loose very shortly, so I decided to have a quick shower and change into my night uniform. It was now about 1700. I decided to take the ship's car and proceed to Fleet HQ and sort out what was required of the ship. Just then the ship's executive officer arrived in his car, and I explained the situation, so we both went in his car to get a briefing.

* * *

Like Bill Ritchie, Brian Swan, Captain, AM RAN (Retired) was also serving on the aircraft carrier HMAS *Melbourne* when Tracy hit Darwin. He remembers Christmas 1974 in Sydney as a fine warm day. He and his wife, Gloria, spent the morning opening presents with their children. Swan had bought a crystal radio set for his son and, in trying it out, he caught the news that a cyclone had hit Darwin. He says, 'Not thinking anything more about it, the family went to my sister's house at Winston Hills for Christmas lunch, which was most enjoyable. But somehow, I didn't feel right, and I didn't have any alcohol at all. I just felt a bit off.'

Swan and his family left his sister's house at about 7 pm and headed home. His unease proved to be well founded. As soon as they arrived at their house, their neighbours came running out, telling Swan he had been recalled to the ship to help Darwin. He changed into his uniform and Gloria and another neighbour drove him to Garden Island to rejoin *Melbourne*.

* * *

Vice Admiral David Shackleton AO RAN Rtd was a lieutenant at the time, with the role of commander's assistant on HMAS *Melbourne*. He was 'tucking into a great lunch' at home with his then-wife, Dianne, and his mother, Elsie, when he received a phone call. Lieutenant Tim O'Sullivan, a duty officer of the day on *Melbourne*, told him that all leave was cancelled and the ship would be sailing the following day to lead the Navy relief team.

Shackleton says, 'We didn't know how long we would be away for – one of the less endearing implications of being in the Navy. Hence, we accelerated the lunchtime festivities and I packed for going away to sea again, about two months in advance of when we had planned.'

* * *

Many others had similar experiences. Lieutenant Richard Usher was the acting executive officer of the Perth class guided missile destroyer HMAS *Brisbane* during the pre-Christmas leave period. The XO was LCDR Tony Townsend, who was on leave in South Australia. Richard and his wife, Bronwen, were living in married quarters in North Ryde at the time. Bronwen remembers:

> We'd been keeping an ear out for news from Darwin. My brother and his wife lived there and the little news we'd had was not good. But it was Christmas, and we had a house full of people to celebrate it with a slow Christmas lunch. By 2 o'clock the Christmas pud was almost demolished when the phone rang.

The officer on duty (OOD) of *Brisbane* had rung, advising Usher there was a 'problem' in Darwin and that the commanding officer required him to return to the ship immediately. His family was still eating their Christmas lunch, but Usher got into uniform, told his guests not to leave as he would be back soon, and drove to Garden Island, arriving there before 3 pm. Usher says:

> The CO, Captain Mike Hudson, was already on board and we had a discussion about recalling the ship's company. We set about doing just that, and over the next few hours a number of officers and sailors returned to the ship. The engineers and electrical personnel worked to put back items under maintenance, and ready the ship for sea.

After Usher's cursory goodbye before he left home, Bronwen recalls, 'What we didn't realise was that was the last we'd see of him for the next seven weeks.'

* * *

Les Renfrey, Commander RAN (Retired), was the weapons electrical engineer officer (WEEO) on HMAS *Brisbane* in

Sydney when the cyclone hit Darwin. He, his wife and their twins were preparing to visit close friends that afternoon when he received a call from the OOD of HMAS *Brisbane*, saying, 'a disaster had befallen Darwin and that the RAN might be called on to provide help'.

Renfrey explains, '*Brisbane* was at Garden Island for an assisted maintenance period and about half the ship's company was on leave.' Renfrey was asked to provide a telephone number where he could be reached for the rest of the day. He says:

> We had only been at our friends' place for an hour or so when the feared telephone call came. So, it was home, pack a bag and be driven to the ship, to work all night: closing up machinery which had been opened up for inspection, maintenance and repair, and doing functional checks to enable a report of 'W.E. department ready for sea'.

* * *

Simon Gourlay, Lieutenant Commander, RAN, Legal Officer (Retired), was an able seaman writer working in the pay office at the Naval College in 1974. He finished his posting there before Christmas and went on leave to Hobart, his hometown. After that, he expected to be posted to HMAS *Melbourne* in late January. He says:

> The news about Cyclone Tracy hitting Darwin on Christmas Eve 1974 was splattered all over the TV the next day. Bearing in mind back then it was black and white TV for us anyway. It was also on the radio. Oh my god, once you heard it, you couldn't picture what it was like until you actually saw the footage on TV and in newspapers, which flowed a few days later.

Gourlay remembers Christmas in those days as being very family oriented. His grandparents, parents and sisters would gather at his parents' house in Rosny, near Hobart. His father would hold an open house on Christmas morning so that their neighbours and friends could drop in, have a drink and exchange Christmas greetings. Gourlay recalls that Christmas wasn't the same as usual as in the back of their minds they were thinking about Darwin, the loss of life and wondering how they were all coping up there.

Likewise, Lieutenant Commander David John, RAN (Retired) also has vivid memories of that time. When Cyclone Tracy devastated Darwin, he was a petty officer medical posted to the Daring class destroyer HMAS *Vampire* (11), which was in refit at Garden Island in Sydney. He and his wife had spent two years at HMAS *Melville* in Darwin, before moving to Sydney in January 1974, and had endured a couple of cyclones in their time up there. Those cyclones had simply passed by, so when they saw on the news on Christmas Eve that a cyclone was heading toward Darwin, they weren't concerned.

John says, 'Little did we know what was in store for Darwin that Christmas night.'

They and their seven-month-old daughter celebrated Christmas with another Navy couple and their two small children at Belmore, 'blissfully unaware of what had happened in Darwin during the night. It was not until around 1000 hours that our friends turned on the radio for the news. When we heard the devastating news, we were gobsmacked to say the least.'

* * *

The recall of naval personnel to their ships was broadcast by radio and television stations all around Australia. Naval personnel also received the call by word of mouth and phone calls from neighbours, friends and police. They responded promptly.

During the night of Christmas Day, as naval personnel began to arrive back in Sydney, arrangements were made for transport from Mascot Airport to the railway station, Garden Island and HMAS *Kuttabul*. *Kuttabul* played a major role in the assistance provided to Darwin. Initially, the base provided administrative support to process naval personnel being deployed to the various ships taking part in Operation Navy Help Darwin. Next, *Kuttabul* provided accommodation for evacuees arriving from Darwin. Later, the base assisted in processing and dispatching naval personnel who arrived in Sydney too late to join their ships before the fleet sailed to Darwin.

In a truly impressive response to the recall, 2020 of the 2700 sailors on annual leave returned to their ships prior to sailing, many of them having travelled from country areas as far away as North Queensland and Western Australia. These were the days before computers, the internet and mobile phones. The grapevine was extremely effective. Volunteers from other ships and shore establishments also reported for duty and were allocated to vacant billets in the ships that were nominated to sail for Darwin.

One of those recalled to service was Tony Vine, a nineteen-year-old marine technical propulsion (MTP) on the Tide class replenishment oiler HMAS *Supply* (A195). He was promoted to kellick just before sailing for Darwin.

'Christmas 1974 was my fourth in the Navy,' he says, 'and like the previous three I expected to have it at my parents' home at Chatswood in Sydney.' When Vine joined the Navy as a fifteen-year-old, he hoped to serve on a sleek modern destroyer or even a frigate. He admits that *Supply* was not the mighty warship of his dreams, but he considered her a great ship, 'despite her portly appearance and slow pace'. She had so far taken him to Gladstone, Brisbane, Melbourne, Hobart, Wellington and Auckland.

A week before Christmas 1974, *Supply* had a change of command, with Captain John Stevens taking over from Commodore Bruce Loxton. Vine describes Loxton as 'an old school officer', whereas Stevens was 'a sailors' captain, friendly and engaging'. He says that Stevens would prove to be the ideal man for the task ahead and quickly earned the respect of the ship's company.

Vine had spent the morning of Christmas Eve on board the maintenance ship HMAS *Stalwart* (D215) in the engineering workshops, machining up a new shaft for a large valve critical to the operation of the lube oil system on *Supply*'s main turbine. He expected the job to fill in the couple of working days between Boxing Day and the New Year. He spent Christmas with his family, going to Midnight Mass, opening presents early at the insistence of his ten-year-old sister, followed by Christmas lunch.

'The TV and radios stayed off,' he says, 'as we settled into our normal Christmas routine.' At the time, his other sister Pat was dating Dave Mitchell, who she would later marry. Mitchell was a shipwright who had been in the intake behind Vine at *Nirimba* and who was posted to the Daring class destroyer HMAS *Vendetta*. Mitchell, who lived in the nearby suburb of Lindfield, came over at 2 pm to wish Pat's family a Merry Christmas. When he arrived, he rushed in to tell them that he'd just heard a news flash on his car radio 'about a bad cyclone up north and that the RAN had lost a patrol boat'.

The family's radio was immediately turned on and they heard the announcer saying, 'The crews of the following naval ships have been recalled: HMAS *Melbourne*, HMAS *Stalwart*, HMAS *Supply*, HMAS *Vendetta*, HMAS *Brisbane* and HMAS *Hobart* …' Mitchell and Vine stared at each other, stunned. Then Mitchell said goodbye and headed back home to pack his bag.

Vine was back on board *Supply* within the hour and changed into his overalls. He went across to *Stalwart* to finish his machining job. He said:

> I had clear and direct instructions from the Chief Tiff that the shaft was to be finished and the valve installed by early the next morning. It was finished by 0200 and deemed functional. While it was not the work of art I had envisaged, the chief was happy, and he allowed me to race home to pack the rest of my kit with instructions to be back on board by 0800. Not much sleep was had that night.

Vine describes the next thirty-six hours as a blur. There was a lot of work to be done including the reassembly of two of *Supply*'s three main boilers, which were in pieces for cleaning. The fleet maintenance party also had stripped down both diesel generators for deep maintenance, along with the ships' freezers, and these needed to be reassembled urgently.

> The technical sailors on board bounced from job to job, as one task was completed another would appear. Fuel and water lighters arrived alongside, and the ship settled deeper and deeper into the water until the boot topping was a good six inches underwater. Meanwhile, a constant stream of sailors began to return from across the city and the country. The enormity of the task of getting such an important ship back to sea quickly became apparent.

Instead of sailing out on Boxing Day, *Supply*'s sailing time was pushed out to midday on 27 December.

* * *

When the motor vessel *Nyanda* in Darwin Harbour was finally able to establish communications with the outside world, Captain Eric Johnston used her to send his first Situation Report to Sydney at 5:52 pm EST on Christmas Day. By early evening, the extent of Darwin's devastation was apparent, and the urgency of preparing the ships for sea was upgraded. Richard Usher of HMAS *Brisbane* says, 'By mid-evening, the whole Garden Island dockyard resembled a working day, with many civilian workers assisting naval staff. Cranes were working. Lights were on everywhere.'

As for HMAS *Melbourne*, she 'was in a maintenance period and was at forty-eight-hours-notice for sea,' says Brian Swan, a lieutenant at the time. Swan describes the ship as 'a big Meccano set that had been pulled apart'. Forty-eight hours was the maximum time allowed for equipment that had been removed for maintenance to be refitted and for the ship to be ready to sail in an emergency.

When the ship had gone into maintenance, most of the crew took leave and travelled home to different parts of the country. Once on board, Swan was tasked with ringing crew members on the recall list, though being Christmas Day, he 'didn't have much luck'. This is where the media response – radio and television – was crucial in getting the message out to sailors around the country to return to their ships. Swan remembers Captain Adrian Cummins of HMAS *Brisbane* being escorted through the city at speed, with lights flashing and sirens sounding, by the local police to Garden Island.

Bill Ritchie was also busy on board *Melbourne*. He and the ship's executive officer had been to Fleet HQ for a briefing. 'On return, I did a quick inspection of all the ship's living spaces and found there were no problems in any area and then sat down to my dinner,' he says. Soon after, he was interrupted by the quartermaster telling him a truck had arrived and was on the wharf alongside the ship, waiting to unload cargo for Darwin. The truck needed to be unloaded urgently as the driver had more cargo to be collected. Ritchie noticed 'a short, slim sailor' approaching the gangway and asked him if he'd been drinking. When the sailor replied no, Ritchie then asked if he was a qualified forklift driver. When the sailor affirmed that he was, Ritchie sent him to acquire a dockyard forklift to unload the truck. The sailor returned with a particularly large forklift and, according to Ritchie, looked 'like a pimple on a pumpkin' inside the forklift cabin.

The aircraft carrier's engines were down, and the ship was only using power provided by the dockyard. When Ritchie climbed up to the flight deck, the crane driver told him that, because of limited power, he could only raise the topping lift, turn left or right, or lift the load. The two ship's lifts were not operational, also due to the power shortage, so the ship's own forklifts were unable to be moved to the flight deck. When the loads reached the flight deck, they were placed as far as possible from the ship's side. Gradually, a semicircle of items was placed on the flight deck, and with each incoming load, the space available was decreasing rapidly.

Once more people arrived on board, full power was soon provided, which enabled the two lifts to be raised. The sailors could use their own forklifts to move items from under the crane onto the lifts and take them down into the hangar. At this stage, *Melbourne* was unable to use a Bailey bridge which would have allowed trucks to drive right up onto the flight deck where they could be unloaded using the ship's forklifts.

With full power restored and the ship's lifts and forklifts operational, Ritchie was able to hand over the responsibility for loading and attend to the myriad of other matters that needed attention. He says the rest of the night passed in a blur and it was well past midnight before he took to his bed. He was up early the next morning to ensure that the two operations rooms and equipment that he was responsible for were in good working order and ready for sea.

* * *

Once good communications were established with Canberra by 8 pm CST, NOCNA issued a series of situation reports, specifying the damage to naval assets and giving an overview of the damage sustained by the city itself. Operation Navy Help was underway. The aim was to provide support to Darwin, to assist in the rehabilitation of the city, and to generally provide maximum assistance to Darwin's people.

The three main supply branches for the Navy – general stores, victualling stores and machinery spares – swung into operation as soon as their heads of department were advised about the task force being organised to provide Darwin relief. Late in the afternoon of Christmas Day, the senior officers of each branch called in storemen and drivers for duty. There had been no warning of such a huge demand for supplies, and no information about exactly what stores would be needed. Senior stores staff went through the storehouses at Randwick, selecting general victualling and machinery stores, which they assumed would mostly likely be needed in Darwin. The experience and knowledge of stores staff contributed greatly to the largest supply task undertaken by the stores section since the Second World War.

The stores staff were advised which ships were available for loading. They then arranged with the Department of Housing and Construction for the supply of building materials and other relief stores from the NSW state emergency services and various municipal council depots.

Roger Fairbairn, the duty officer at FOCEA Headquarters, Sydney, spent the day in frenetic activity, doing his best with only four blackboards and three phones to coordinate the provision of supplies to the fleet. He says:

> Those actions took possibly most of the daylight hours of the first day, and I decided I would stay in my new operations centre. I have forgotten the name of the officer who was scheduled to relieve me; however, whoever it was joined me putting details onto the blackboards. I had arranged for the wardroom stewards to bring over

> sandwiches, coffee and pillows and we slept on the floor, as I remember.

HMAS *Melbourne* could not be shifted south to enable the Bailey bridge – a portable truss bridge connecting the wharf to the ship – to be used, so stores were unloaded at the fitting out wharf. After unloading their cargo, the trucks immediately returned to Randwick for reloading.

Sydney's Garden Island dockyard was reported to have an almost wartime atmosphere as the fleet was prepared and mobilised. Stores personnel continued to work through the night, selecting and readying stores and equipment for dispatch to the task force readying at Garden Island.

The duty watch of HMAS *Melbourne*, under the control of Lieutenant Commander William Ritchie, used the ship's crane to load the material as fast as it was delivered. Around 100 tonnes of stores had been loaded by dawn on Boxing Day.

CHAPTER 16

THE DEPARTURE OF THE FLEET

Thursday 26 December 1974 – Boxing Day

Richard Usher phoned his wife, Bronwen, from HMAS *Brisbane* sometime between midnight and 1 am on Boxing Day morning. He told her he would not be coming home and would need all his kit, so she packed a bag of uniforms, toiletries and so forth but, to Richard's regret, no camera. When Bronwen arrived at the dockyard at about 2 am to deliver her husband's bag, she found a 'frenetic level of activity':

> Nothing was standing still. Lights blazed, vehicles moved every which way and people were striding about in a very purposeful way. The contrast to the usual laidback, 'give us a minute, love' attitude was certainly stark. The 'buzz' was palpable.

This was the last time the husband and wife would see each other until Usher returned to Sydney in mid-February. Usher's only regret is that he left home in such a hurry on Christmas afternoon that he did not say goodbye to their young son. He was busy and doesn't remember much of the early hours of Boxing Day morning at Garden Island, 'apart from everyone working, including civilians, and everywhere was a hive of activity'.

Brisbane's CO called regular meetings with his senior officers for updates on their preparations for sailing. As an acting executive officer, Usher's main concern was personnel and having enough specialists to fill key roles. Officers, senior sailors and junior sailors trickled back to the ship through the night and Usher received a list of those present and the positions that still needed to be filled.

At about 11 am on Boxing Day morning, Usher went ashore to the parade ground on HMAS *Kuttabul* where a large number of officers and sailors had gathered. 'Many were not attached to any particular ship but had answered the general recall,' he says, and 'some were actually men who had deserted the Navy months or years earlier'. Some of *Brisbane*'s key personnel had still not returned from leave so Usher picked out a small squad of specialist sailors and marched them back to the ship.

* * *

Everybody had a story to tell about how they heard about the recall and their efforts to return to their ships. Two stories about sailors in his division stick in Usher's memory:

> The first was a young ordinary seaman RP who was riding his motorbike Newcastle way. Some long way north of Sydney he was pulled over by the police for speeding. He explained he was a sailor returning to his ship, and with that, the policeman switched on his blue flashing lights and escorted my sailor at speeds of up to 140km/h all the way back to Garden Island.
>
> The other was a Western Australian sailor. He turned up at Perth airport and was given VIP treatment by TAA staff and flew back to Sydney – first class – on the 'midnight horror'. I think there were many sailors at the airport, and TAA offloaded civilians to make way

for sailors. At Sydney, TAA had arranged transport to Garden Island.

HMAS *Brisbane* filled up rapidly with returning personnel and volunteers from other ships and shore bases. Stores included twelve outboard motors, nine Hercules aluminium boats, diving gear and large quantities of potatoes, flour and other food, and stores. The monumental effort had *Brisbane* ready to sail by 11 am on Boxing Day. Sailing, however, was delayed.

* * *

Roger Fairbairn, the junior officer working at Naval Headquarters on Boxing Day, recalls, 'The stewards kept bringing something to eat and a steward or two were assisting me in running hand messages to and from Garden Island personnel who were now opening the boat pound and other facilities.'

He was gratified that on the second day, a sense of order was shaping up. He had a discussion with the CO of *Waterhen* about items held in the naval store facilities at Piermont and Randwick that could be needed in Darwin. They chatted about the landing craft alongside and their immediate capability to be stocked and sent to Darwin. They authorised everything they could that would fit on the landing craft. These included building materials, tents, clothing, vehicle parts, tyres and batteries, small generators and pumps, kitchen utensils, water containers and other items that they thought might be useful.

The organisation was ramping up, and Fairbairn appreciated having more personnel around to share the load. When WOC Roberts arrived, Fairbairn brought him up to speed with all the pertinent information noted on the blackboards.

In addition, he says, 'Carleen Angel, a senior admin in my office, came back from leave and she took on the task of typing memos and instructions that we were now sending left, right and centre. Two of the other girls, who usually worked in other parts of the headquarters, also joined in these tasks in due course.'

Naval personnel continued to flood back to Sydney to rejoin their ships, and Ansett Airlines were asked to provide buses to transport sailors from Mascot Airport to Garden Island. The buses were provided along with police escorts. The Chief of Staff, Commodore Dollard, had been contacted and Rear Admiral Dovers returned.

The stores sections worked nonstop during Boxing Day, procuring stores and equipment, which were promptly delivered to the ships at Garden Island Dockyard. The Army also sent equipment with a police escort from Randwick to Garden Island for shipment to Darwin. Fairbairn says, 'There is no doubt that as the situation developed, the naval-stores folks did a magnificent job and worked in a highly flexible way to cut short normal processes.' The fleet was quickly stored and made ready to sail.

The Bailey bridge was connected to the flight deck of HMAS *Melbourne* sometime after dawn on Boxing Day, and the flow of stores on board increased significantly. Unfortunately, some trucks and semi-trailers were unable to use the bridge, which complicated loading somewhat. Even so, it was estimated that over 300 tonnes of general stores for damage repair in Darwin was loaded onto *Melbourne* by 2 pm that day.

David Shackleton returned on board *Melbourne* early on Boxing Day to find the ship 'embarking all manner of stores it was thought might be needed for helping the stricken city'. Though they'd seen the 'terribly bad state of devastation' in Darwin on TV, 'we really didn't appreciate just how bad it was until we arrived,' he says. At the time, Shackleton was a lieutenant, who worked as commander's assistant for the executive officer, Commander George Halley, organising and administering the daily routine of the ship. He describes it as a great job, and one he did in addition to his other roles in operations room watchkeeping and air controlling.

He was assisted by Chief Petty Officer Quartermaster Gunner (CPOQMG) Bert Kindelan, who Shackleton describes as 'a jovial man who epitomised what a good chief petty officer should be – a mentor and sage sceptic with a great sense of humour and always immaculately dressed'. Kindelan would advise Shackleton 'on all things to do with organising duty rosters, watch bills [a list of all members of the ship's company, divided into watches], stores parties and all and sundry required to get 1000 or more members of the crew to the right place at the right time'. On Boxing Day, volunteers were joining *Melbourne* from other ships to be part of the rescue team. With missing crew members of their own, their task was particularly difficult. They worked and reworked the watch bills half a dozen times.

Another unforeseen problem was the difficulty of contacting civilian contractors on public holidays. The current contractor for fresh fruit and vegetables could not be contacted on Christmas Day. Instead, at 7:15 am on Boxing Day, *Melbourne*'s order was phoned to the home of Mr Goon of R. Goon & Son, the only known source of supply who could be contacted immediately. Mr Goon rang back at 8 am to report on what fresh fruit and vegetables he could supply. These included cabbages and lettuce being cut and brought in by market gardeners from Horsley Park and other districts.

By 8:45 am, Producers & Providers Ltd, the official contractor for fruit and vegetables, was contacted and asked to supply the rest of the requirements for *Melbourne* and the other ships of the task force. The meat contractor, Presto Smallgoods Ltd, also made their services available. The civilian contractors delivered meat, butter, cheese, fruit and vegetables directly to the ships.

817 Squadron and its Wessex helicopters had disembarked from the ship prior to Christmas and were at Naval Air Station

(NAS) *Nowra* when Tracy hit. HS 817 Squadron personnel were among those recalled to duty and they returned to Nowra. The squadron promptly flew to Sydney, joined HMAS *Melbourne* and set to work immediately.

In addition to the work involved in loading stores onto the aircraft carrier and organising enough personnel, David Shackleton says that quite a bit of *Melbourne* needed putting back together before she could sail. This included the notoriously poor air-conditioning system, as well as the evaporators which converted saltwater to fresh by boiling the saltwater and capturing the cooled condensate as fresh water. He says that the evaporators normally couldn't produce the amount of fresh water the ship needed, and it wasn't unusual for the ship to be on water rations. These issues would be particularly problematic in the tropical north.

'The ship was a hive of activity,' adds Brian Swan of HMAS *Melbourne*. 'No-one slept. Throughout the night, the engineering staff worked flat out, and steam was raised.' The plan was to sail immediately after the yachts competing in the Sydney to Hobart race had sailed on Boxing Day. 'And we did,' he says. 'It had taken less than thirty-six hours with a vastly reduced crew, but we were ready to sail at 1300 on Boxing Day.'

The order came to remove the Bailey bridge but nothing happened. Bill Ritchie remembers thinking, *The bloody Dockies, at it again*, but the problem turned out to be a blown fuse in the shoreside crane's electrical system. The fuse was soon replaced, the Bailey bridge removed and *Melbourne* was ready to set sail. After a truly mammoth effort, at 5 pm on Thursday 26 January 1974, the aircraft carrier HMAS *Melbourne* and the guided-missile destroyer HMAS *Brisbane* sailed from Garden Island, Sydney.

HMAS *Melbourne*, commanded by Commodore GR Griffiths, carried the commander of the naval task force, Rear Admiral DC Wells. HMAS *Brisbane*, commanded by Captain

MW Hudson, was under orders to proceed at high speed to Darwin. HMAS *Stuart* (48), the River class destroyer escort under the command of Captain ET Keane, sailed at midnight on 26 December and would overtake *Melbourne*.

* * *

Friday 27 December 1974

Following media broadcasts that all naval personnel were to report back to their ships and establishments, particularly in Sydney, Petty Officer Medical David John reported for duty on HMAS *Vampire* on 26 December. While watching television coverage of some of the badly injured Darwinians being flown south, John recognised a police officer who had been in his Rugby League club in Darwin in 1973. John visited the injured police officer in the Prince of Wales Hospital and the officer talked about the terror he and his family had been through. His family were safe, but he had been hit by flying sheets of iron and debris and said he was lucky to be alive. John says, 'This gave me the resolve to try and get back to Darwin to do my bit, if possible.'

After reporting at HMAS *Vampire*, PO John attended the office of the fleet medical ward master, Fred Ashmore, to offer his services where possible.

'I advised Fred that I had been in Darwin for the two years 1972–73 and, as I knew the area very well, it would be advantageous for me to be part of the relief medical teams,' says John. Ashmore agreed and John was seconded to HMAS *Stalwart*. The next day, 27 December, John bade his wife and baby daughter farewell and joined *Stalwart*. The ship's company had been working hard for two days loading some 300 tonnes of disaster-relief stores.

The destroyer tender *Stalwart* was an Australian-designed and constructed escort maintenance ship and proved invaluable during Operation Navy Help Darwin. Roger Fairbairn, who also served on *Stalwart*, says it was 'a wonderful ship' and perfect to send to Darwin because it could provide electrical power for up to four ships, and had large cool rooms and dry stores. He believes *Stalwart* was the most important ship to get to Darwin.

* * *

HMAS *Brisbane*, *Melbourne* and *Stuart* had left the day before, but other ships were still filling with stores. On 27 December, Tony Vine of HMAS *Supply* was sent to the dockyard naval stores with the engineering storeman to get as many nuts, bolts, gasket material, spanners and other tools as they could. He says they'd been given carte blanche to take anything they needed and filled their trolleys to the brim while the civilian storeman 'madly' wrote down everything they chose.

They had to be back on board by 2 pm, and suddenly realised it was already 1:45 pm and they were on the other side of the island from their ship at the west dock wall. As they 'bolted out the door of the store', the storeman yelled at them not to remove the trolleys from the store, but they pretended not to hear him. They arrived at the wharf just as the crane was in the process of lifting the gangway. The bosun, spotting them lumbering along the wharf, detailed off a swarm of sailors who raced down the gangway, picked up the overloaded trolleys and carried them back up. 'Two days later, the trolleys were ceremonially buried at sea off the Queensland coast,' Vine adds with a laugh.

While HMAS *Supply*'s official report of proceedings for December 1974 says *Supply* 'sailed at 1400 on Friday 27 for Darwin', Vine points out that it makes no mention that it did so on only 'two main boilers with no serviceable diesel generators'. Of *Supply*'s 240 crew, all but twenty-one had made it back and the shortfall was made up by sailors seconded from ships in the major refit and shore establishments.

Patricia Collins as a new WRANS recruit, mid-1973.
Source: Author's collection

Patricia Collins on Darwin Wharf, 1974.
Source: Author's collection

Official class photo of WRANS Recruit Class 140.1973, February 1973.
Source: Author's collection

Navy Headquarters destroyed by Cyclone Tracy, 25 December 1974.
Source: Kenneth Doust Collection/Library & Archives NT

Coonawarra Navy base and houses destroyed by Cyclone Tracy, 25 December 1974. Source: Kenneth Doust Collection/Library & Archives NT

The wreck of the navy patrol boat HMAS *Arrow* after Cyclone Tracy.

Source: Royal Australian Navy

HMAS *Attack* driven aground by the force of Cyclone Tracy, 27 December 1974.

Source: Vic Sumner/*The Sydney Morning Herald*

HMAS *Advance* surrounded by debris following Cyclone Tracy, December 1974.

Source: Albert A. Dixon Collection/Library & Archives NT

Michael and Christine Pilkington's house in Darwin after Cyclone Tracy. The graffiti on the wall reads 'PILKINGTONS ALL OK'. Source: Christine Pilkington

Cyclone Tracy damage in the northern suburbs of Darwin.
Source: Coburg Collection/Library & Archives NT

Locals on night patrol, protecting vehicles from looters in the aftermath of the cyclone, 30 December 1974. Source: Rick Stevens/*The Sydney Morning Herald*

Navy crew of HMAS *Vendetta* clean up Darwin, 1975.
Left source: NAA: A6180, 28/1/75/9
Right source: NAA: A6180, 28/1/75/13

Navy helicopter landing. HMAS *Melbourne* in the background.

Source: NAA: A6135, K17/1/75/49

Cyclone relief in Darwin.

Source: NAA: A6135, K17/1/75/56

Patricia Collins, 1976. Source: Author's collection

The fleet maintenance party (FMP) got to work rebuilding the generators.

The ship's freezer was also still in maintenance, so a large, refrigerated container full of meat and frozen vegetables had been secured to the boat deck aft. Of greater concern, however, considering they were heading to the tropics, were the many sections of the ship's air-conditioning system that were also down for maintenance. Despite these problems, *Supply* set sail in company with *Hobart* and *Vendetta*, with *Stalwart* following suit four hours later. The task group was now at sea and proceeding to Darwin.

* * *

The advance guard, HMAS *Brisbane*, *Melbourne* and *Stuart*, which sailed on Boxing Day, were followed a day later by the destroyer tender *Stalwart* commanded by Captain IM Burnside, the fleet oiler *Supply* commanded by Captain JD Stevens, the guided-missile destroyer *Hobart* (39) commanded by Captain PR Sinclair and the Daring class destroyer *Vendetta* commanded by Commander AG Ferris. They were all loaded with more relief supplies.

Stalwart rendezvoused with *Hobart*, *Vendetta* and *Supply* for the trip north along the east coast for Darwin. The RAN's five landing craft heavy (LCH) – *Balikpapan* (L126), *Betano* (L133), *Tarakan* (L129), *Brunei* (L127) and *Wewak* (L130) – had also been deployed from HMAS *Moreton* in Brisbane, Queensland.

In his subsequent report in March 1975, Rear Admiral Neil McDonald, Flag Officer Commanding East Australia Area, wrote:

> I cannot speak too highly of the enthusiasm and application displayed by members of the storehouse staff at all levels. In the first rush of activity on Christmas and Boxing Day, most were on their feet for sixteen

> to nineteen hours continuously. There was an air of urgency and cooperation, the likes of which I have not experienced since wartime, and the staff virtually worked themselves to a standstill.

After the task force's two squadrons had sailed, all further stores were airlifted to Darwin by RAN or RAAF aircraft. Returning and volunteer naval personnel reported to HMAS *Kuttabul* in Sydney and were then dispatched to Townsville by air, to meet up with the ships on their voyage north.

* * *

Back at FOCEA Headquarters, the work continued unabated. Roger Fairbairn comments, 'After about forty-eight hours, Navy Office advised that Major General Stretton had been appointed to coordinate the Darwin support response and we stepped back from lead control. I didn't know at that time that the National Disasters Organisation existed.'

Fairbairn says FOCEA Headquarters 'started getting signals from Canberra from this organisation and the orders were confusing and off the mark. Some of the things that were coming through did not actually fit in with reality.' He adds:

> I had a negative view of the Canberra organisation, initially because it was Army focused and they really didn't seem to understand Navy type things. But anyway, I think largely the Navy did its own thing by itself because in those days the Army would have had very little ability to get their people up to Darwin.

FOCEA staff continued to provide administrative support for several weeks to supply the equipment transported to Darwin for naval use. Everything was better organised, but Fairbairn remembers the first forty-eight hours as very difficult.

It was just totally off the wall, the whole thing. We didn't even have a map room. I had charts, but we didn't have any instructions on procedures to follow in emergencies or disasters. Nothing about what you would do to organise a response. They had some processes set up at Fleet Headquarters, but we didn't have anything. Nothing.

CHAPTER 17

THE SHIPS' JOURNEY NORTH

HMAS *Melbourne*

Most of the ships in the fleet had departed Sydney without their full complement of personnel. They had volunteers from shore bases and other ships but were still below strength. HMAS *Melbourne* was particularly short-staffed. Bill Ritchie recalls helicopters being deployed to embark members of the ship's company and other personnel as the ship made its way to Brisbane. The helicopter squadron also brought more pallets of relief equipment.

With about one-third of the regular crew on board, the available personnel strived to make up the labour shortfalls, with many taking on tasks outside their usual duties and working longer hours. Lieutenant Brian Swan, for example, had to steam the forward boilers, which was usually a petty officer's job.

HMAS *Melbourne* hove to off Townsville on 29 December, where they embarked 230 recalled personnel who had been unable to reach her in Sydney before she sailed. The 'ferry load of sailors' who greeted them in Townsville, says Swan, 'were immediately pressed into the steaming watches'. *Melbourne* also took on more disaster relief and victualling stores. Captain Eric Johnston had supplied information about the damage to the city of Darwin and its naval assets and set out a provisional list of requirements. Detailed planning commenced during the passage of the ships to Darwin.

Every member of the ships' companies was asked to list his skills and qualifications apart from those obtained during his naval service. An incredible diversity of talent was revealed, with many having civilian qualifications previously unknown to their superiors. HMAS *Brisbane*'s crew, for example, included: former carpenters; bricklayers and plasterers; powerhouse, telephone exchange and house-wiring electricians; butchers; pipe-layers and plumber's mates.

Melbourne's captain appointed Brian Swan as works manager based on Swan's work on Manus Island, and he was included in the headquarters group. They made plans en route to Darwin, says Swan, for managing the task ahead.

HMAS *Brisbane*

The guided-missile destroyer HMAS *Brisbane*, also known as the 'Steel Cat', was commanded by Captain MW Hudson. According to a February 1975 article in *Navy News*, *Brisbane* proudly took the honours of being 'first to sail, first to arrive, first to land and last to leave'.

Two-thirds of the ship's company had returned, and their numbers were boosted by volunteers from other ships and establishments, so, although still understaffed, the ship could be run in three watches.

When *Brisbane* set out from Garden Island, Sydney Harbour was still packed with small craft that had stayed after watching the start of the Sydney to Hobart yacht race. Richard Usher says, 'HMAS *Melbourne* and *Brisbane* were given a rousing farewell. Once outside the heads, HMAS *Brisbane* was dispatched at full speed. This was thirty-two knots.' Because of the high speed, the CO ordered double officers of the watch (OOW) on the bridge. The more senior officers, including Usher, kept OOW duties and the junior officers, who were the usual OOW, supported them. For Usher this role, combined with his acting executive officer role, was onerous. He worked long hours with little rest, but heartily appreciated having two officers on duty on the bridge when the vessel was travelling at such a pace. Usher recalls, 'This was a time before GPS, so fixing the position of the ship was manually done, and at that speed, things happen quickly. Navigating through the Great Barrier Reef at thirty-two knots is not for the faint-hearted!'

The water became quite shallow in their approach to Townsville, with extra hazards to navigate. For the first and only time in Usher's naval career, he experienced 'a shallow-water effect'. When the gap between the ship's hull and the seabed is restricted, the propellers sink lower into the water to get more thrust. This causes the bow to rise which in turn reduces the speed of the ship.

As a midshipman, Usher was asked in a test once about shallow-water effect but didn't know what it was then. He says the Navy is normally cautious about going too fast in shallow water, so the effect is rarely encountered. When the CO realised that they were experiencing shallow-water effect, Usher and the other officers were called to the bridge to have a look. 'We reduced revs to twenty-eight knots, and the ship's bow lowered to the correct position, but our speed stayed the same,' says Usher. Approaching land, the ship once again showed signs of shallow-water effect, so they reduced the revs to twenty-four knots for the ship to right itself a second time. Usher says this happened a few more times that day – a rare occurrence for any naval ship.

During the voyage, the ship was constantly updated, not only on the situation in Darwin, but also about what stores would be available for loading in Townsville. The majority of the ship's company who had missed joining the ship in Sydney were also waiting in Townsville. To Usher's relief, this included the executive officer (XO).

Until then, the planning for arrival in port and the loading of supplies was Usher's responsibility. He was advised that there would be 'several large trucks mainly with food and immediate survival equipment to support the small remaining population in Darwin'. They would also embark about forty Dan Buoys which are small temporary marker buoys, each with a lightweight flagpole.

Storing a ship involves getting men to line up along the gangway between the truck and the ship. 'Boxes are passed chest to chest,' says Usher, 'so there is effectively a conveyer-belt operation.' As the ship would be in port for less than three hours while it refuelled, Usher ordered three gangways to be made available. One was for use by personnel embarking and disembarking and two were for 'store ship' – the carriage of cargo onto the ship. About thirty sailors were detailed for each gangway.

The magnificent people of Townsville, however, had other ideas. 'There were about fifteen trucks lined up waiting for us, and hundreds of Townsville folk to greet us,' Usher recalls. 'As the first gangway was craned into position, the first truck pulled up, and men and women of Townsville swarmed in to start unloading. They formed the human chain to convey the stores up along the gangways – not allowing my sailors to do the work.' Usher rapidly reorganised the stores parties to stow everything below decks. The fifteen trucks were unloaded, and all stores were on board within forty minutes.

The first man up the personnel gangway was the XO. Usher recalls him saying, 'Well done, Usher. You carry on as XO until special sea dutymen are called. I don't want to interfere with any of your planning.' After those few hectic days, Usher says that was the most wonderful thing anyone had ever said to him.

After loading stores, refuelling and welcoming more personnel on board, HMAS *Brisbane* departed Townsville and continued her urgent sail to Darwin. Usher found it exciting to keep watch on the bridge while steaming at high speed within the Great Barrier Reef. Navigation had to be exact, and there was no room for error. They fixed their position every ten to fifteen minutes, day and night. In order to avoid small vessels, both bridge staff (OOWs and lookouts) and operations-room staff had to be vigilant.

At least after Townsville, Usher could relax a little as he was no longer responsible for running the ship and could revert to his normal duties of running the operations room. Meanwhile, the medical staff were busy ensuring everyone was up to date with vaccinations. Detailed planning was underway for their arrival off Darwin in the early hours of 31 December.

* * *

Les Renfrey was a lieutenant commander, the Weapons Electrical Engineer Officer (WEEO) of *Brisbane.* He also remembers the increased workloads on the trip north. 'Briefings were given on the Darwin situation and what our task would be on arrival. Inventories were taken of stocks of hand tools that would be needed and any materials that might be of use in making repairs.' In rare quieter moments, stories about answering the recall signal were shared among the ship's personnel. Several sailors from Western Australia told of what happened when they fronted up at Perth Airport. Airline staff boarded a plane warming up for take-off and pulled passengers off to make room for them.

One officer, hastening to the ship, was pulled over for speeding while crossing the Harbour Bridge. When the police officer heard why, he provided an escort, complete with lights and sirens, to the dockyard.

The Australian population assisted in many unexpected and deeply gratifying ways: police officers provided escorts, market gardeners picked fresh vegetables, Townsville citizens loaded naval vessels, people sent clothing, children donated their pocket money. The people of Darwin were the grateful recipients, directly or indirectly, of all this generosity.

HMAS *Stalwart*

The escort maintenance ship, *Stalwart*, commanded by Captain IM Burnside, sailed for Darwin on 27 December after thirty-six hours loading stores and emergency materials. As with all the ships sailing north, the ship's company was surveyed to identify non-service skills relevant to the task ahead. In civilian life, before joining the RAN, many sailors had been bulldozer drivers, heavy-equipment operators, licensed house wirers, electrical linesmen, auto electricians, motor mechanics, fencers, tilers, carpenters and joiners. All these skills would be in demand in Darwin.

Downtime was scarce and precious, and celebrated in the best traditions of the service. David John, who assisted with the running of the sick bay, fondly remembers celebrating New Year's Eve in the senior sailors mess as *Stalwart* rounded Cape York. 'The petty officer cook ensured they had plenty of appropriate victuals,' he says, 'to go with the limited supply of alcohol that was legally permitted.' John continues:

> When the latter had disappeared, the mess members brought out their personal 'sea stocks' from their lockers, which enabled the party to continue for a while longer. Once these were depleted, the victualling senior sailor advised that he just happened to have a case of communion wine in the store which was unlikely to be used by the chaplain. Accordingly, that wine was drawn from the victualling store and appropriately disposed of in true Navy fashion. Whether the wine contained any alcohol was irrelevant as the word 'wine' was enough to attract the interest of the members.

HMAS *Supply*

The Tide class replenishment oiler *Supply* had sailed from Sydney at 2 pm on Friday 27 December. In Townsville, the bulk of cyclone stores were moved to the ship by jackstay transfer.

Tony Vine says, 'Sadly, the ships were also supposed to pick up a large amount of Army bedding and stores; however, the Army officer in charge in Townsville refused to release them, arguing that the "requirement for them in Darwin had not been confirmed".' The remainder of the trip was relatively quiet, Vine says, except for two events:

> On the afternoon of New Year's Eve, we passed two landing craft, HMAS *Betano* and HMAS *Balikpapan*, struggling gamely toward Darwin at twelve knots. That evening off Cape Grenville, we sat on the poop deck having our beer issue while we passed a merchant ship heading south. The ship was brightly lit and the sounds of music and singing drifted across the water to us.

The commanding officer of the landing craft heavy HMAS *Betano*, Lieutenant Graeme G Rogers, noted in his report of proceedings that, 'Although it is realised by all LCH crews that a maximum speed of ten knots is not overly fast, the frustration with this slow speed on this occasion and in these circumstances was really felt as all the other units of the naval relief task force sped by.'

CHAPTER 18

THE ARRIVAL OF THE FLEET

The first naval ships to arrive in Darwin were the guided-missile destroyer HMAS *Brisbane* and the survey ship HMAS *Flinders*, which reached Darwin waters at 9 am CST on 31 December. *Flinders*, commanded by Lieutenant Commander JJ Doyle, immediately check-surveyed the approaches. Their arrival was not only morale boosting for the remaining civilian population of Darwin but was greeted with palpable relief by the naval population as well. We were shabby, sad and worn down with clean-up duties. The sight of the initial shore parties in crisp white uniforms was the first sign that help had arrived. Our tattered ranks were soon bolstered by hundreds of fit, strong young men, disciplined in the service and ready to take on the task of cleaning up Darwin.

NOCNA Captain Eric Johnston was also relieved to have assistance and reported later that 'the augmentation' of his staff with the fleet commanders 'removed a great burden from myself and my remaining staff officers'. As for the fleet personnel, they provided 'clearing teams for the town in general'. He added that they also manned the headquarters on shore and assisted in repair work on RAN facilities.

As civilian Barry Chester – Lieutenant Richard Usher's brother-in-law and Darwin resident – attests, the clean-up in Darwin was well underway by the time the first naval ships arrived. While Chester didn't experience the cyclone personally, he returned the following day to assess the damage to his house and electronics business. The industrial area where his business was located had fortunately not sustained too much damage from the cyclone. With the water back on in that area, 'many men camped in the surviving sheds, putting up as best they could with the heat and rudimentary sanitation, and maybe an improvised shower once in a while'. Chester admits:

> We were luckier than most: not only had the shed that housed our electronics business survived the cyclone, but we even had air conditioning courtesy of our own generator, a massive old Ford V8, which was soon handling half a dozen fridges and freezers and a washing machine, plus a tangle of extension cords to surrounding sheds. After the day's work was over, my business premises soon became a popular choice for dropping in for a bit of cool air and a beer or three.

As for the rest of Darwin, 'some basic post-Tracy order was slowly being restored by the tireless efforts of a handful of survivors,' Chester says. By the time help from interstate arrived, water and phones were restored in some areas and the major roads were mostly cleared of debris. Communal kitchens were set up to feed the people who were working long hours in the stifling heat and humidity, and 'drinking plenty to compensate'. He remembers the all-pervading stench of rotting food, and the abandoned cats and dogs scavenging through the debris that still littered the ground. It was the Army, Chester says, who initially tackled 'the huge and awful job of disposing of this noxious mess'.

HMAS *Brisbane*

Richard Usher, on HMAS *Brisbane*, recalls that when the ship arrived off Darwin in the early hours of 31 December, the CO dispatched all senior officers to points of importance in Darwin to assess what needed to be done. The navigating officer took the ship's boat, laden with Dan Buoys, and an old-fashioned hand lead and line, to chart the passage into Darwin Harbour. It was not known if the cyclone had altered the seabed in any way. He took the boat along the normal entry route, marking the way with a Dan Buoy every 200–300 metres or so, having ensured there was sufficient depth for the large vessels, HMAS *Melbourne* and HMAS *Supply*. Meanwhile, the CO had gone ashore to call on NOCNA, Captain Eric Johnston.

Usher was the direction officer, responsible for the ops room, RP sailors and aircraft control. He was sent to the airport to assess radar equipment, Tacan (a navigational aid system for aircraft that measures bearing and distance from a ground beacon), and other equipment. The airport was 'a complete mess with some light aircraft upside down'. The radars and the Tacan had been demolished. Usher was requested to provide what assistance he could to help aircraft safely enter and exit Darwin airspace.

He returned to the ship and immediately set up the operations room for aircraft control duties and arranged watchkeeping staff accordingly. His deputy direction officer was Peter Stehn, and together they ran a round-the-clock aircraft advisory service.

When the navigating officer returned to HMAS *Brisbane*, the ship was able to steam into Darwin Harbour and drop anchor. Usher sent out a NOTAM (notice to airmen) saying what the ship's Tacan frequency was and their position, offset from Darwin Airport (bearing and distance).

Usher then mounted a 24/7 service bringing aircraft in and out of Darwin Airport safely. That lasted for about a week until the RAAF was able to set up a portable radar and Tacan, and their air traffic controllers could resume operations.

Usher also took the opportunity to meet up with his brother-in-law, Barry Chester, who knew the Navy was on its way. Chester was looking forward to seeing Usher, his sister's husband, who soon appeared at Chester's workshop, 'brisk and confident in his immaculate Whites'. Usher's appearance was uplifting, Chester says, to the 'group of untidy and preoccupied men in various stages of chronic tiredness'. Usher's white uniform 'proclaimed a new team had arrived' that was 'ordered, competent and reliable'.

Chester's business partner, Brian, was an Englishman who'd served on the lower deck in the Royal Navy and had a very low opinion of officers. Brian was immediately suspicious of Usher who spoke with what Chester called 'a clipped British accent'. Brian was not entirely comfortable sharing a congenial beer or two with the RAN officer until Usher mentioned he too had started on the lower deck and risen to 'upper yardsman' in the Royal Navy. Brian's attitude, according to Chester, 'changed from truculence to respect in an instant'.

* * *

Les Renfrey had accompanied HMAS *Brisbane*'s CO and heads of department when they did their initial tour of Darwin. He found the place weird and quite eerie. They saw the effects of Tracy's wind speeds – up to 215 km/h – in the devastation around them, but now there were 'only light breezes bringing smells of insecticide'. Crop-duster aircraft had sprayed insecticide to control the insects attracted to 'decaying flesh and rotting vegetation'. There were no flies and the shore party noted all the trees stripped of leaves.

Accompanied by NOCNA, Captain Eric Johnston, the initial shore party was taken on tour by police through the suburbs of Darwin. Among the memorable sights they encountered was a car in the pool of the Travelodge Motel. Close by, at the main marina, small craft were smashed and piled into heaps. Another sight was the martini-glass-shaped water tower that had sustained severe damage about halfway up the column. A crumbled vehicle lay at the bottom. As for the houses, which predominantly sat on stilts, most upper levels were destroyed. The occasional bath remained intact with the pipe to the showerhead bent at about forty-five degrees. Opening refrigerator doors, Renfrey says, 'was a particularly nasty experience. Food, especially seafood, which had been in an enclosed space for seven days or more in searing heat had a stench hardly imaginable.'

After their initial tour of the city with local police, HMAS *Brisbane*'s commanding officer and his heads of departments organised clean-up teams who would go ashore daily. A skeleton crew would remain on board to 'steam auxiliary' on one boiler for electrical power, keep the ship safe and be available to make preparations in case the ship had to sail at short notice.

The marine engineer officer (MEO) had construction experience and was an obvious choice for going ashore to liaise with local authorities and direct efforts. His deputy stayed on board as did Les Renfrey, the head of the other engineering department. The three junior weapons electrical engineer ordinary seamen (WEEOS) were assigned to clean-up teams.

The aircraft carrier *Melbourne* and the River class destroyer escort *Stuart* arrived at 12:30 pm on 2 January 1975. The destroyer tender HMAS *Stalwart* arrived at midday on 2 January 1975. On 3 January, the fleet oiler HMAS *Supply*, the guided-missile destroyer HMAS *Hobart* and the Daring class destroyer HMAS *Vendetta* arrived at 9:30 am. The landing craft *Betano* and *Balikpapan* arrived in Darwin from Brisbane on 4 January, and the remaining landing craft *Brunei*, *Tarakan* and *Wewak* arrived on 13 January.

HMAS *Melbourne*

When the aircraft carrier HMAS *Melbourne* was still 150 miles out on her approach to Darwin on 2 January, the

Fleet Commander Rear Admiral HD Stevenson and his headquarters group embarked a helicopter and viewed the state of the devastation by air. Stevenson reported:

> The first impression of Darwin was one of virtual hopelessness. It is extremely doubtful if any personnel had expected the widespread and almost total destruction of such large areas, particularly in the northern suburbs of Nightcliff, Rapid Creek and Casuarina which made up the Naval assistance area. Realisation instilled an even greater sense of urgency in the efforts to get relief work started quickly.

The headquarters group met with Mayor Harold 'Tiger' Brennan, the chief of police and NOCNA Captain Eric Johnston. Johnston, who was known as 'Big E', went on to be promoted to commodore and subsequently became the Administrator of the Northern Territory. Brian Swan remembers the police wanting sailors to be armed to prevent looting and to shoot dogs which were hunting in packs looking for food. Admiral Stevenson, however, was having none of that. His personnel 'were there to help the city and people of Darwin, and if there was a problem with law and order that was a civil matter'. Swan adds that no-one 'in their right mind would let armed sailors ashore'.

The sailors' duties would be confined to several key areas. The first was in the suburb of Nightcliff, making houses watertight where possible for remaining residents. The second task was to restore key buildings and services to accommodate contractors and other tradespeople who were arriving in Darwin. The third goal was to get the patrol boat fleet serviceable again. NOCNA prioritised clearing the slipway and inspecting the boats for damage.

Also in the headquarters group was Bill Ritchie, newly appointed as HMAS *Melbourne*'s disaster relief coordinator.

Their aims, he says, were to make houses safe, remove debris and place it close to the road for trucks to come and take away. If any valuables were found, they would be placed in a safe area. As most destroyed houses were built on stilts, a safe place could be found under the house or what was left of it. Teams were briefed that if any alcohol was found it was to be put in a safe place. Any rotting food was to be removed and placed adjacent to the road. Engineering and electrical sailors were to be employed to repair equipment and re-establish power lines and make them safe.

With all parties agreeing on the priorities, the headquarters group flew back to *Melbourne* to promulgate the necessary orders to the other ships and the aircraft carrier's personnel.

Each ship's company was then briefed on their duties. Bill Ritchie says, 'They were made aware that Darwin's population of women and children had been flown south. It was emphasised how they would feel if it was their destroyed home and people were taking their possessions.'

The HS 817 Squadron's Wessex helicopters on board *Melbourne* made an immediate impact on Operation Navy Help Darwin. The squadron was able to conduct search and rescue sorties in areas that had been inaccessible by road, and enabled evacuations to be carried out rapidly, even over great distances.

CHAPTER 19

THE FLEET AT WORK

HMAS *Brisbane*

HMAS *Brisbane* moved to her assigned anchorage just off the Darwin wharves at 1 pm on 31 December, and the plans that had been made during the voyage were immediately put into action. Boats were lowered into the water and working parties proceeded ashore.

The priority was to establish a headquarters at Admiralty House for the commander of the Australian fleet (COMAUSFLT) when he arrived on HMAS *Melbourne*. The next, and longer-term, focus was to start clearing up the suburbs that had been assigned to the Navy: Casuarina, Nightcliff and Rapid Creek. These three northern suburbs were designated the sole responsibility of the naval relief teams for search, rescue, clearance and safe storage and recording of any valuables found. On their first day on the job, forty-four tonnes of stores were landed by helicopter and 250 personnel were landed to carry out a half-day's disaster relief work.

A skeleton crew stayed on board at all times. Chefs continued to feed everybody, and the engineers and greenies (electricians) kept the ship at operational readiness. One way or another, everybody contributed to the relief effort. Once RAAF air traffic control officers resumed operations, Lieutenant Richard Usher was able to stand down his radar plot sailors who'd assisted him and Peter Stehn, and they were able to join the daily shore parties. Usher felt much pride in going ashore to assist in the mammoth task of cleaning up. He says, 'The Navy was there to help. We gave everything and took nothing.'

Every day, several hundred men went ashore. As the Navy had their own rations, water and accommodation, and could even dispose of their own garbage, they were not a drain on the city's limited resources, making what they were doing, in Usher's view, a really worthwhile task.

The self-containment of the Navy is something Brian Swan, of HMAS *Melbourne*, also points out. From a civilian perspective, says Swan, the self-sufficiency of the 'constantly busy' Navy was certainly appreciated. Their men didn't need any support and didn't put any load on the almost non-existent services ashore. Additionally, Barry Chester says that they and the Army were 'well organised, cheerful and ready to turn their hands to anything, including some seriously unpleasant clean-up jobs'.

There was, however, a shortage of suitable equipment and heavy machinery for demolition and debris removal at first, making the clean-up work slow and physically arduous for their teams. Richard Usher recalled the lack of appropriate clothing and PPE for the sailors. They wore navy-blue shirts and shorts, and either boots or sandals. Working ashore six days a week, almost everyone got sunburnt. Despite the conditions and lack of PPE, Usher says morale was high and 'almost everyone was really happy' to do what was required. There were few defaulters and sailors did what they were asked without question. Les Renfrey agrees, noting those who volunteered to work on Sundays were joined by members of the skeleton crew, including himself, who had previously been left on board. He says the skeleton crew relished 'the chance to contribute in a more direct way'.

* * *

HMAS *Melbourne*

HMAS *Melbourne* arrived in Darwin Harbour early on 2 January 1975. David Shackleton, commander's assistant on HMAS *Melbourne*, recalls seeing 'the normally lush mangroves near the harbour approaches stripped bare'. It looked, he says, 'like the kind of effect an airburst by a nuclear weapon might cause'. There would be little time to contemplate the scenery, however, and the men were soon put to work. *Melbourne*'s arrival enabled 1200 naval personnel to be landed from the ships every day. Some were airlifted ashore by helicopters and others by ships' boats and landing craft, which departed at 6 am every day.

The shore coordination efforts, as mentioned, were run by Captain Eric Johnston.

Shackleton's job, along with his assistant CPOQMG Bert Kindelan, was to help administer the ship for Commodore Guy Griffiths. They also arranged the 'work parties that went ashore by boat in the early hours and came back to a stinking hot ship in the afternoon'. The relief parties worked from sun-up to sundown. They were ferried ashore by boat every morning and then bussed to their allocated neighbourhoods for clean-up duties. They spent their days clearing the debris of wrecked houses and felled trees – tasks far removed from their normal duties.

Using aircrew briefing facilities, Shackleton says they would plan the day before for the personnel needed for the following day's tasks. Shackleton, Kindelan and two of his staff were 'responsible for nominating and allocating people to *Melbourne*'s working parties and keeping track of who did what'. The planning that had taken place on the trip north paid dividends. Shackleton reflects on the fact that they worked out and recorded all the data with paper and pens. He concedes that computers and spreadsheets would have made the process much easier and more efficient.

Once the organisational side of things was under control, Shackleton assigned himself to a working party.

He recalls one day, 'amid the oppressive heat and humidity, dirt, sweat and toil', a woman appeared at their headquarters in Nightcliff and asked for help. The headquarters was set up on the concrete base of a wrecked house. There were stumps holding up a floor and 'nothing else left of the house to speak of'. The woman explained that her husband's pay packet was on the bedside table when the cyclone hit and was hoping someone would help her find it. Shackleton says they looked at each other doubtfully, thinking there was absolutely no hope of finding a pay packet amidst the rubble. Nevertheless, half a dozen sailors took some tools and went to what was left of the woman's house. After about half an hour of searching, they found her husband's pay packet. 'Sailors love being able to do things with chainsaws and power tools that chop stuff up,' Shackleton says, 'and miracles can happen.'

* * *

'When anchored in the harbour,' says Brian Swan, 'everyone who was not actually on watch would be ashore doing their bit.' Team leaders, says Swan, reported to shore headquarters at Captain Eric Johnston's house each day to receive their tasks. Meanwhile, as Bill Ritchie recalls, the teams on *Melbourne* were called at 4:30 am so they could shower, have breakfast and collect a packed lunch and water, before assembling in the hangar. Here, the groups were briefed on their task for the day, as provided by headquarters ashore. It was reiterated that they must not keep any items they found when clearing houses and rubble. This, 'it was made clear', included alcohol.

Helicopters would then transfer everyone ashore from *Melbourne*'s flight deck. Each person was checked off against a list and would be checked off again once they'd returned to the ship at the end of the day. They were also searched for any items they may have taken from shops or houses they'd been working in. Ritchie says there was only one occasion when someone tried to bring stolen items back to the ship.

On shore, Lieutenant Brian Swan's job was to manage about 450 tradespeople doing a vast array of work. He commandeered a Holden utility to drive around in and fuelled it from a service station which the Navy had also taken over.

Heat, high humidity and daily rainfall made conditions ashore uncomfortable. Life wasn't easy for the shore parties even when they were back on board *Melbourne*. 'It was seriously hot,' Shackleton recalls, and 'the air-conditioning refused to cooperate' despite the best efforts of the engineers. Shackleton's cabin was 'under the flight deck and that big slab of steel acted like a huge heat sink'. Despite the breeze, which would sometimes pick up after nightfall, it never cooled down to a comfortable level. Many men slept outside on the deck.

* * *

Around 600 naval personnel worked daily at the especially hard-hit suburb of Nightcliff. A helicopter pad and a sub-headquarters were established there to facilitate the work details. Working in their assigned areas, Lieutenant Brian Swan and his team found their ingenuity challenged when faced with many new and unexpected situations. Sam Ashton, a warrant officer electrician, was one such member of Swan's team – 'a truly wonderful guy' who 'could do anything', according to Swan. Out in a utility truck one day, they came across a generator with a flat battery, making the sewerage system it was supposed to be pumping unserviceable. 'Sam looked around and found some cable lying in the street, which had come down in the cyclone,' recalls Swan. Ashton instructed Swan on 'which wire to put where and hey presto' – they managed to jump-start the generator, get it running and restore sewerage to the area.

Warrant Officer Ashton's other big achievement was getting the air-conditioning plant of one of the big hotels in the city going again, as the area was largely uninhabitable for tradespeople. Electricity had been restored thanks to electrical linesmen from South Australia and Queensland who had streamed into Darwin to assist in the recovery efforts. The air-conditioning plant was located in the building's basement which had been flooded. The plant's motors were wet and inoperable. Swan says Ashton's answer was to slide the motors up a man-made ramp from the basement to the street and then to hail a passing helicopter to lift the motors onto the hotel roof. Once there, the motors were dismantled and the windings and rotor assemblies were laid out to dry in the sun. By the afternoon, Swan noted the 'resistance readings were within tolerance'. The motors were returned to the ground by helicopter and slid back into the basement. Sam Ashton restored the air-conditioning and tradespeople could be accommodated at the hotel.

Swan recalls another incident in Nightcliff when teams of sailors were set to work emptying Christmas roasts out of ovens. They dug holes in the backyards and buried the roasts along with the rotten contents of refrigerators and freezers.

'Sounds good, doesn't it?' Swan says. 'But what do you do with sixty tonnes of rotten prawns?'

That's exactly what they came across on the waterfront. Northern Research had a big freezer where the catch from their trawlers was stored. Tracy cut power throughout the city and Swan says, 'The prawns stunk!'

That ingenious team got a portable generator, re-froze the prawns and, when they were frozen solid, loaded them into a tip truck, took them to a swamp and dumped them. 'At least,' says Swan, 'it removed the health hazard.'

I wish we'd thought of doing that sort of thing. Clearing out fridges and freezers was easily the most sickening task I undertook and there would be many who'd agree. I think that using portable generators to refreeze rotten food in places like supermarkets and butchers prior to disposal should be enshrined in every disaster management plan devised from here on.

Richard Usher, from HMAS *Brisbane*, also recalls the 'overwhelming smell and feel of rotting food'. He and his team wore anti-gas respirators when cleaning out the cold store at Woolworths one day, which reduced the effects of the smell, but Usher will never forget 'the maggots!'

Captain Eric Johnston made 'special mention' of this particularly unpleasant part of the clean-up operation with reference to the commendable efforts of the fleet band master and his musicians. As musicians, the men needed to 'protect their hands from damage and were not permitted to carry out normal clearance work'. They were, instead, organised into hygiene parties, tasked with the disgusting job of emptying domestic fridges and freezers. The musicians recoiled with revulsion at the stench and texture of the rotting food. They suffered frequent and severe bouts of vomiting. Johnston notes that 'the situation deteriorated' when the men were tasked with clearing the cold and cool rooms in the supermarket.

On the lighter side, at one house in Nightcliff, a clean-up team found 'a most interesting array of dildos, ticklers and pornographic colour slides,' says Swan. Unsurprisingly, word of the illicit cache spread very quickly, and street maps were marked with its location. The 'equipment', as Swan calls it, was never claimed.

Beer, or the lack of it, was also a serious issue. 'Hot weather and the nature of the tasks made for thirsty sailors,' says Swan. Commander George Halley convinced the ship's captain that the sailors should receive a double beer issue upon their return each day. Getting the brewery and pubs open for business was an integral part of returning the city to normal, but Swan says a team could not be left at a pub or brewery for too long. The temptation of being surrounded by so much alcohol could be a recipe for trouble. A decision was quickly made to rotate teams through the pubs and brewery every three days and then return them to other clean-up duties.

Elsewhere in town, the Travelodge, one of the city's largest hotels, had opened its bars, and its sodden carpets had been removed. However, other Darwin services and infrastructure faced barriers to reopening. The international airport, for instance, was unable to reopen until the quarantine station was habitable. The leprosarium, which still had patients in it, also needed its power and sanitation restored.

Tracy had ripped the roof off the building that housed Northern Territory archives, resulting in important archival material and papers being drenched. During the clean-up operations, Bill Ritchie was summoned to the Disaster Relief Headquarters to meet an archivist. The gentleman requested the assistance of a small work party to pack important archival papers which were a sodden mess. Ritchie put together a team who helped the archivist pack the papers so they could be flown to warehouses in Brisbane and Sydney to be dried out. 'Imagine the surprise,' Ritchie says, 'when the boxes were opened in the south, and all the papers were completely dry. Miracle or karma.' Due to the change of air pressure in the aircraft, the moisture in the papers had apparently evaporated. 'You learn something every day,' adds Ritchie.

* * *

Many RAN members involved in the clean-up of Darwin after Tracy had served in the Vietnam War and others had previous experience dealing with disasters. Simon Gourlay, a former lieutenant commander and retired legal officer, worked on two disasters in one month.

At 9:27 pm on 5 January 1975, *Lake Illawarra*, a bulk ore carrier, was travelling up the Derwent River when she collided with several pylons of the Tasman Bridge. A large section of the bridge deck collapsed onto the ship and into the river below. Twelve people were killed; seven of them were crew members and five were the occupants of four cars that fell forty-five metres (150 feet) after driving off the bridge.

Gourlay, an able seaman writer at the time, was at home in Hobart on leave for Christmas. He was at the Elwick Drive-In on the western shore when the film was stopped, and an announcement was made on the screen about the bridge collapse. Before he left home that evening, Gourlay had told his mum, 'If the movie's no good, I'll be home early.' His family lived at Rosny on the eastern side of Hobart, just 500 metres away from the bridge, and his mother panicked at the news, wondering if her son was on his way home or not.

In the early hours of the morning, a fourteen-man Navy clearance diving team flew to Hobart to assist the Tasmanian water police in the recovery of the vehicles that had driven off the bridge.

The next morning, Simon Gourlay was back home with his family when he received an urgent telegram to say that his posting to HMAS *Melbourne* was postponed, his leave was cancelled, and he was to report to HMAS *Huon*.

He caught a ferry across to the western shore and proceeded to HMAS *Huon* to assist with the Tasman Bridge disaster. Gourlay was soon on the police boat *Vigilant*, tied up to one of the pylons, which he describes as 'quite a scary experience'. Engineers hadn't inspected the bridge at that stage and big water pipes were banging against the pylons, raising fears of further collapses. Gourlay didn't know whether he was going to get home that night.

He worked with the Navy divers and the police for a week or so. The *Navy News* later published a photo of Gourlay with the Navy divers in a small open boat underneath the bridge.

Then, Gourlay was given another posting: instead of joining HMAS *Melbourne* as previously instructed, he was to join

HMAS *Brisbane* in Darwin. His first reaction was apprehension. He'd seen snippets of the devastation on TV and some photos in the newspaper but didn't really know what to expect, or what he would be assigned to do when he got to Darwin.

He left Hobart at about 6 am and flew to Darwin via Melbourne and Adelaide on a Fokker Friendship (a turboprop aircraft). He arrived in Darwin at about 8 pm that night. He says, 'I remember getting off the Ansett plane and walking through the airport, seeing that it was badly damaged but wasn't completely demolished. It wasn't until you got out to the 'burbs that you saw the complete devastation. It was just amazing. It's beyond belief.'

His journey wasn't over yet. He still had to get out to his ship, anchored in the harbour. 'It was quite a rough trip in a little boat, and I was in my Whites,' he says. It was compulsory for Navy personnel to wear our formal uniforms when travelling, but 'they weren't very white by the time I got to the ship!'

When he arrived on board, all the bunks were full. He was posted to *Brisbane* as a writer to do pay and accounts, so for weeks he simply slept on a camp stretcher in the pay office until he eventually ended up with a bunk down in the aft seaman's mess.

He was promptly assigned to a working party for the clean-up. As *Brisbane* was anchored mid-harbour, the crews had to be picked up by Navy landing craft which would arrive early each morning. They were given 'sandwiches and lots of cordial and juice', says Gourlay, which were loaded aboard the landing craft, before they took off to start their daily clean-up routine. Once ashore, they were allocated an area of responsibility and required to clean up a certain number of houses per day.

Their job was to remove hazards, rubble and debris that needed to be carted away and dumped. Gourlay remembers being astonished at finding freezers that had been blown down the road for 200 or 300 metres. As well as the freezers full of food meant for Christmas celebrations, they also came across toys, which Gourlay found 'heartbreaking'. Most property was beyond repair but valuables were handed in, bagged and taken to headquarters at the end of the day.

Gourlay remembers steel telegraph poles that were bent horizontal to the ground – something he found hard to comprehend. He was also amazed by how lucky most people were. 'Some went to their bathrooms and survived,' he says, 'others went somewhere else in the house and the bathroom was demolished.' Gourlay finds it incredible that so many people survived the cyclone.

Clean-up efforts were hampered by the persistent rain, which turned the ground and rubbish to slush. He ended up 'having three tetanus injections after treading on nails and stuff like that'. With the lack of PPE, Gourlay wore boots, short pants and a handkerchief around his neck. Many sailors didn't even have hats and they were all continually getting sunburnt.

'It was a vicious cycle,' he says, 'of burning, peeling and getting burnt again.' When working near a local swimming pool, they were at least able to have a refreshing dip every now and then.

Like all the members of the relief crews, Gourlay was going ashore early every morning, spending the day cleaning up and returning at the end of the day, 'pretty much exhausted'. And his work wasn't finished once he was back on board his ship. As a writer, he had to do the pay and accounts which were done manually every fortnight. Naval personnel were paid in cash, and he remembers doing the payroll until midnight, or sometimes to one or two o'clock in the morning. He would sleep in the pay office, pack up his camp stretcher in the morning, go ashore and work on the clean-up, return to the ship and work on the payroll. 'It was all good fun,' he says. 'We just got on and did the job, as we were expected to do. And came away relatively unscathed, I think. We did a pretty good job.'

The heat and humidity of Darwin in the wet season, plus the lack of sewerage systems and drinking water, took a heavy physical toll on the naval personnel. The work we were doing, often sorting through the remains of people's possessions, was at times dispiriting. A heartening sight was the many naval officers performing manual labour alongside the sailors. The front pages of southern newspapers featured a photo of Commodore Guy Griffiths, captain of HMAS *Melbourne*, stripped to the waist and carrying a large baulk of timber.

The phenomenal Navy effort in the early days of the new year, along with more people arriving each day to help with the rebuild, saw Darwin beginning to get on its feet again. As more heavy equipment was transported to Darwin from the southern states, the rate of progress also improved.

Along with the clearance and salvage work, up to 500 personnel a day were tasked with skilled duties such as the repair of Postmaster General (PMG – now known as Telstra and Australia Post) facilities, and general electrical wiring.

'Electricity was being restored apace and some shops were opening,' says Swan. He also believes 'the sounds of helicopters flying over the city each morning' was reassuring to the remaining civilian population. The personnel of the ships in the fleet proved themselves invaluable in the clean-up effort and rebuilding of the city.

The clean-up of the city and the repairs were heavy, hard work, often gut-wrenchingly disgusting, but were more achievable than mending the people who suffered from the cyclone. HMAS *Brisbane*'s WEEO Les Renfrey went ashore to work on the clean-up on the first Sunday of the operations and recalls:

> … going to a warehouse where I ran across the husband of one of my wife Ros's school mates. He had been sent

> by his company to rally the troops and see what he could do with the mess. He said he had found the staff in a bad way, pointing to one man still sitting in a dark corner, whimpering gently.

The best efforts of Renfrey's friend were futile in easing the man's broken spirit. Many people who went through the cyclone reached the limits of their physical and emotional strength sooner or later. Darwin city recovered and went on to grow and thrive. Many people did not.

* * *

The patrol boat base at Stokes Hill Wharf fared badly in the cyclone. Most of the roofing had blown away and the interior had suffered widespread water damage. The Port Authority's slipway was damaged in the cyclone and was unusable. A slipway is a ramp leading down into the water. Ships can be moved out of the water and onto the slipway on rails for repairs, and then slipped back into the water. While on the slipway, boats are stabilised on cradles which keep them upright and immobile. The Port Authority's slipway was needed urgently for repairs to be undertaken on the Navy's patrol boats.

The prawning vessel NR *Liverpool*, despite being secured on a cradle in the Port Authority's slip, had been knocked over during Cyclone Tracy and was leaning at an angle of forty-five degrees against a concrete pier. The slip cradle had also been lifted away from its rails and there was concern that it had been damaged.

'Some of the more interesting tasks we did were a real challenge,' says Brian Swan. Few tasks could have been more challenging than clearing and repairing the slipway, but he says, 'We did it.'

Lieutenant Terry Knowles, *Melbourne*'s shipwright officer, says Swan, 'did a great job in righting the trawler', returning 'it to the water using gravity and some muscle power on the slip's out-haul wire'. The inspection of the cradle revealed no damage so the go-ahead was given to slip HMAS *Attack* for repairs.

After HMAS *Attack* had been moved off the rocks and berthed alongside the wharf, Bernie Kloss and the rest of the ship's company worked hard to clean up the patrol boat and keep it watertight.

'Eventually,' he says, 'it was deemed safe enough to get around to the slipway.'

We became used to improvising, making do and thinking outside the box. For instance, Wran Joan Czarnecki found herself unexpectedly assisting at Stokes Hill Wharf when she had some time off.

The sailors from one of the damaged patrol boats needed to drive down to the boat on the wharf and sail her around to the slipway. They then needed someone else to drive the car to the slipway for them. Joan volunteered but drove 'in first gear all the way', worried the police would pull her over because she only had a motorbike licence.

With no electricity, Brian Swan and his team had to get *Attack* up onto the slipway. Lieutenant Swan improvised by using a D6 bulldozer – a procedure he'd seen performed in New Guinea. Once the patrol boat was safely up on the slip, Swan reported the news to NOCNA. Captain Johnston was glad to hear it and later reported the shipwrights had used 'considerable initiative and improvisation' to return the prawning vessel to an upright position and then to slip *Attack*.

Kloss says the shipwrights took a couple of days to weld up the largest holes in *Attack*'s hull and make her watertight again. By the time *Attack* came back alongside the wharf, HMAS *Stalwart* was berthed on the other side of Stokes Hill Wharf, unloading much-needed building materials and stores.

The patrol boats *Advance* and *Assail* were slipped in quick succession for inspection and repair. Later, it was discovered that the concrete-clad steel columns supporting the overhead gantry rails above the thirty-tonne slip were cracked.

* * *

Stalwart, *Hobart*, *Vendetta* and *Supply* had arrived in Darwin Harbour on 3 January, joining the small armada of RAN vessels already there.

On entering the harbour, Petty Officer Medical David John says, 'We simply could not believe our eyes at the destruction that we could see from the ship.' Once ashore, they would soon discover it was even worse out in the suburbs. Most of the ships were anchored in the harbour, but HMAS *Stalwart* tied up at Stokes Hill Wharf to offload the disaster relief supplies. Their teams would add an extra impetus to the efforts of personnel from *Melbourne* and *Brisbane* and supply some electricity to the city. After briefings that afternoon, John was tasked with setting up a first-aid post (FAP) with Leading Seaman Medical (LSMED) John Bellwood and Able Seaman Medical (ABMED) Peter Lonsdale the following day.

On 4 January, the three medics went ashore with the ships' clean-up crews and boarded a bus to be transported to their allocated clean-up area. To John's delight, his assigned area was Nightcliff, his old neighbourhood. The medics were dropped off at Nightcliff AFL Oval, and the clean-up crews were taken around the suburbs to commence work. HMAS *Melbourne*'s clean-up crews were being delivered to the oval by Wessex helicopter, so the medics set up their FAP in a corner of the oval, just inside a six-foot-high wire fence, where they would be out of the way of the helicopters but still visible to any casualties who needed assistance.

The medics had no cover and were sitting on the ground out in the sun for the first hour or so. They had no communications and their equipment consisted of one Oxy-Viva Resuscitator, an Army pattern stretcher and various medical kits. A local, seeing their exposed position, provided them with a tarpaulin which he no longer needed himself. The medics stretched the tarpaulin over the corner of the wire fence for some much-needed shade. Sitting on the ground was still not ideal though, so John strolled over to the demolished clubhouse, hoping to find some chairs. In addition to numerous stackable plastic chairs, John found bar refrigerators still full of beer and soft drinks. He carried about half a dozen chairs back to the FAP as the first casualties started to trickle in. John says, 'At least we could now sit them down comfortably while we treated their cuts and abrasions.' He then 'got to thinking' that they could offer the patients a drink to cool them down and organised a courier to send milk cans full of iced water to their post later that afternoon. He then 'got to thinking even more that perhaps the casualties would like something more refreshing than cold water to drink'. To that end, he collected many cans of beer from the clubhouse refrigerators and placed them in the milk cans with the iced water – a move that proved popular with the patients.

Unknown to John, clean-up crews were finding plenty of alcoholic beverages in and around demolished houses. Some of the wine, spirits and beer recovered was being imbibed immediately, while the rest was smuggled back on board at the end of the day. After day two, bag searches were done when crews returned to their ships and the practice was stopped. On their second day at Nightcliff Oval, John recalls 'a big flash black sedan' pulling up near their FAP. Admiral Wells and some other Navy officers and civilian dignitaries stepped out and approached the FAP. They asked John and his team how they were going and how many casualties they'd had. The admiral then asked, 'What's in the milk cans?'

John froze in horror as they were fully stocked with iced water and beer cans. He recovered his voice and answered, 'Simply iced water for the injured.' The admiral took his word for it and left, after which John and his team heaved a huge sigh of relief.

On their third day ashore, the FAP was moved to the same location as the headquarters group. As David Shackleton notes, this was set up under the floorboards of a demolished house. The house was adjacent to the football oval, across from where the FAP had been originally set up.

John says, 'Obviously, our beer provision service came to a grinding halt at that time. Still, it was fun while it lasted.'

The medics didn't have a vehicle to transport patients to hospital if necessary, so John tasked LSMED John Bellwood to check if Nightcliff High School, the local refugee centre, had a suitable vehicle for loan. Bellwood headed off on foot to the school and returned shortly afterward with his chosen vehicle: a red motorcycle.

John stared in disbelief. When he got his voice back, he asked why Bellwood had selected a motorbike. Bellwood suggested the bike could be used to get around the neighbourhood to provide a mobile first aid service. An admirable thought, but Bellwood was the only one who could ride a motorbike and, as John says wryly, 'It did not meet my requirement of a vehicle that could carry a stretcher.' John decided to go to the high school himself to try and obtain a more suitable vehicle.

As luck would have it, John knew the fellow in charge of the transportation area. He was a former chief petty officer electrician who had served with John at HMAS *Coonawarra*. The two chatted for a bit before the ex-serviceman directed John to the carpark on the school football ground and told him to select whatever vehicle he wanted.

'There were a couple of dozen vehicles,' says John, 'of all makes and sizes, which had either been abandoned with the keys still in them or on loan for general use.' John selected a Toyota utility, which enabled him to drive around and see the damage firsthand.

The first place he checked out was the house he and his wife had lived in during their time in Darwin. He then travelled around the suburbs further north of Nightcliff 'where the damage appeared to be even worse'. He visited HMAS *Coonawarra* and dropped into the senior sailors mess right at lunchtime. The damage, he noticed, 'was nowhere near as bad as the coastal suburbs'. By this time, we had already done a great deal of clean-up work at *Coonawarra* and much of the rubble had been removed. John found it 'heartbreaking' to see the magnitude of the damage throughout the city.

John, like everyone I interviewed, only had praise for the clean-up crews:

> They were doing a magnificent job under the circumstances, particularly in the heat and humidity of the tropics. For some, it was heartbreaking to venture into a residence to find children's and other Christmas presents still wrapped and waiting for an owner who was never going to receive them. The crews took great pride in their untiring efforts in cleaning up the city of Darwin.

* * *

On the morning of 3 January, HMAS *Supply* arrived off Darwin, refuelled *Brisbane* and then proceeded to its anchorage in Darwin Harbour. The threat of further cyclones and severe storms was high so, like all the ships in harbour, *Supply* had to remain on four-hours-notice to proceed to sea. This severely limited the number of people who could go ashore. Tony Vine found that he 'was to remain on board for that first week and each night the working parties would return and regale us with stories of what it was like ashore'.

The *Supply* sailors, Vine says, mostly worked 'cleaning up houses at Fanny Bay and Nightcliff'. After breakfast at 5 am, they were transferred to *Melbourne* by boat, then taken ashore by helicopter. Later, as the helicopter route proved inefficient, they were taken directly by boat to *Stalwart*, alongside Stokes Hill Wharf. The men were issued fruit, sandwiches, two canteens of water and a soft, floppy 'giggle' hat, such as the Army usually wore but in navy blue. They returned to *Supply* between 5 and 6 pm, 'exhausted but still enthusiastic about the job they were doing', says Vine. He remembers *Supply*'s commander, John Stevens, grabbing a beer one evening and talking with the sailors about their experiences ashore. Vine describes Stevens as a great leader who was well respected. There was a big boost in morale when the beer issue was raised from one to two each night, with Carlton White, VB or Fosters on the drinks menu.

Vine finally got to go ashore at the end of *Supply*'s first week in Darwin.

'Despite having heard all the stories,' he says, 'nothing prepared me for the devastation.' He describes houses that were 'mere slabs on pillars with small piles of debris, or the skeleton of a wall attached, personal possessions scattered everywhere, already rotten and stinking in the monsoonal conditions'. Like many, Vine remembers 'vividly' the smell of rotting food in fridges that contained Christmas dinners, including seafood. Clearing out one site, the team secured the door of one fridge with a BandIT strap to prevent it from opening before hauling it away. After forty-six years, 'the smell of rotten prawns still evokes memories of Darwin,' he says.

More heartbreaking 'was having to throw out family treasures which were beyond recovery' – items such as clothing and photo albums went straight into the garbage truck. Salvageable items such as ornaments and jewellery were bagged. Their original locations were recorded and they were placed in storage for eventual return to their owners.

This routine was basically the same for the clean-up crews from all ships. The sailors went from house to house, clearing the houses and yards and piling debris on the roads.

Front-end loaders would then come along, scoop up the debris and deposit it into large dump trucks to be disposed of out of town. Refrigerators and freezers were opened, and the rotting food taken out for urgent disposal. Any valuables located were handed to an officer, who recorded the item, its description and where it was found. The valuables were later handed over to the civilian authorities with the relevant information.

* * *

Clean-up crews put in eleven- to thirteen-hour days, Monday to Saturday, for the duration of Operation Navy Help in the heat and humidity of Darwin's wet season. Returning to their ships at the end of the day, sailors were mainly interested in bathing, donning clean clothes, enjoying their beer issue and a hearty meal, and then collapsing into their beds. Even so, many Darwin residents commented on the cheerfulness and work ethic of the sailors and officers as they carried out the heavy work of clearing debris.

CHAPTER 20

BOARD OF INQUIRY

A Board of Inquiry (BOI) was convened on HMAS *Stalwart* to look into the 'circumstances of the loss and damage to HMAS *Arrow*, *Attack*, *Assail* and *Advance* and the casualties resulting therefrom'. Commodore Guy Griffiths, captain of *Melbourne*, was president of the Board of Inquiry.

BOIs are internal Australian Defence Force inquiries, generally conducted to ascertain the facts surrounding an event that may afterward become a matter for investigation by civil or naval authorities. These events may involve accidents, injuries to personnel or damage to an ADF asset. They recommend actions that may be taken in response, including actions aimed at avoiding a recurrence.

Lieutenant Robert Dagworthy of HMAS *Arrow* was apprehensive about giving evidence at the BOI. 'Having lost my ship,' he says, '[and] two of my ship's company, and being mindful of what happened to the captains of *Melbourne* after the *Voyager* and *Frank E Evans* disasters, I believed that my naval career was finished.'

Dagworthy's fears were well founded. In 1964, the aircraft carrier HMAS *Melbourne* and the destroyer HMAS *Voyager* were engaged in night flying exercises. They were in home waters and the conditions were calm but *Voyager* unaccountably turned in front of *Melbourne*'s bow. The destroyer was cut in half and sank. Out of *Voyager*'s ship's company of three hundred and fourteen, eighty-two were killed, including her captain, Duncan Stevens.

Subsequent investigations put partial blame on *Melbourne*'s command team, despite incontrovertible evidence that *Voyager* had performed a dangerous turn directly in front of the aircraft carrier. Two Royal Commissions followed. Captain RJ Robertson and other officers in *Melbourne* were criticised for not questioning *Voyager*'s movements before a dangerous situation developed. Captain Robertson's career was effectively over, and he eventually resigned.

Over forty years later, Peter Cabban, who had been XO of *Voyager* until five weeks prior to the collision, wrote a book, *Breaking Ranks*, casting serious doubts about Duncan Steven's fitness to command, a situation he claimed was well known in naval circles. Robertson, however, had borne the blame.

In 1969, HMAS *Melbourne* was involved again in a frighteningly similar collision while participating in the joint SEATO exercise SEA SPIRIT. In the early hours of 3 June, the destroyer USS *Frank E Evans* crossed *Melbourne*'s bows and was cut in two.

Melbourne had all navigation lights on for safety's sake. When the *Frank E Evans* turned to starboard to cross in front of *Melbourne*, Stevenson sent a message over voice radio from bridge-to-bridge warning *Evans* that she was on a collision course. The aircraft carrier then radioed to *Evans* that it was turning to port and sounded two blasts on its siren. *Evans* then spun hard right under *Melbourne*'s bows. Seventy-four of *Evans*' crew lost their lives, and *Melbourne* sustained extensive damage to her bow section. Search and rescue operations began immediately, and 199 men were rescued.

A joint USN/RAN Board of Inquiry was conducted and held Captain Stevenson partly responsible for the disaster, stating that as commanding officer of *Melbourne*, he could have done more to prevent the collision from occurring. Stevenson subsequently faced a court-martial.

The evidence and testimony given at the court-martial showed that Stevenson had not done anything wrong. His defence team claimed that he had done everything reasonable to avoid the collision and had acted correctly. The charges were dropped and the verdict 'honourably acquitted' was given. Despite this, Stevenson's next posting was as chief of staff to a minor flag officer, seen by him as a demotion. He requested retirement.

Public opinion was that both Robertson and Stevenson were scapegoats. They were not responsible for the tragedies, but their careers were over.

Bob Dagworthy thought he would be court-martialled and would need to find a new career. He believed he'd 'have a blot on [his] copybook from what happened in Darwin', thinking it would be found that he'd 'made mistakes' and would have to 'accept the consequences'.

After Dagworthy was evacuated, he and his wife, Christine, stayed with her parents in Sydney for several weeks. Though surrounded by supportive family, including his mother who lived close by, Christine remembers her husband's 'terrible heartache' over what happened in Darwin. When the BOI convened, he flew to Darwin and stayed with the *Arrow*'s company at HMAS *Coonawarra* while the crew gave their evidence.

* * *

Leading Seaman Bernie Kloss was still in Darwin working on the patrol boat *Attack*, berthed at Stokes Hill Wharf, when the skipper told him to get into uniform.

> 'We've got an inquiry going on in the chapel on *Stalwart* in an hour and a half.'

> I went, 'Me?'
>
> He said, 'Yeah, well, you were chief engineer.'
>
> I was only a kellick! A leading hand. I just found the first shirt I could, crinkled up with no badges of rank or anything. I only had my apprentice cap. Apprentices never ever wore the round hat, always the peaked cap right from the start. The skipper and I both marched over to the *Stalwart* and waited outside the chapel.

An able seaman came out of the chapel and called for the skipper of HMAS *Attack*, Lieutenant de Graaff. Paul de Graaff went in to face the BOI. When he came out, twenty-or-so minutes later, the same able seaman, unsure of who or where the chief engineer was, simply called for the chief engineer of HMAS *Attack* on the night of 24 December to enter. 'So, I waltzes in,' says Kloss. In the room was a commodore, a couple of commanders, a psychologist and a chaplain. 'There must have been about 600 years of naval sea time sitting in front of me – more gold than you could get in the Federal Mint,' he adds. *This could be very interesting*, he thought at the time.

Kloss's unconventional mix of uniform, especially his apprentice cap, caused confusion. He recalls the commodore looking up, checking the apprentice badge on his cap, looking back down at his file and asking, 'Chief engineer? Charge engineer? Who are you?'

'Leading Seaman Kloss, sir.'

'And why are you here?'

'Well, I was given the duties of chief engineer on the night of Christmas Eve by the captain, Lieutenant de Graaff.'

'You were?'

'Yes, sir.'

'Why?'

'Well, the actual engineer was away,' Kloss said.

'Couldn't they find anybody else?'

'Well, I believe at the time, no-one was prepared to volunteer.'

'Oh, well. Off cap! Sit down.'

While the commodore proceeded to ask questions – 'How many manoeuvres did you do during the night?' – Kloss noticed one officer looking at him intently. Eventually, that officer said to Kloss, 'My speciality is psychology. You don't seem too disturbed.'

'No sir, not really, I don't think so,' Kloss said.

'Did the event affect you?'

'It was a bit scary, sir,' replied Kloss.

'A bit scary?'

Other members of the board continued to ask questions, until Commodore Griffiths finally looked at Kloss and said, 'You know, we will have to change some of these regulations because this situation should never ever have arisen. These people are too young to be carrying out these kinds of duties.' Dismissing Kloss, he said, 'Well, thank you very much ... er ... what is it?'

'Leading seaman, sir.'

'Well, thank you very much, Leading Seaman. Dismissed. On cap.'

And off Kloss went.

* * *

At the end of the Board of Inquiry, Lieutenant Bob Dagworthy was called to HMAS *Melbourne* and told to meet with Commodore Griffiths in his cabin, where he expected to be told that he was to be court-martialled.

Commodore Griffiths was at his desk. He handed Dagworthy a signal and said that these were the recommendations from the Board of Inquiry which would affect him. The commodore had highlighted the sections he wanted Dagworthy to read.

'In part,' Dagworthy recalls, it stated that he 'should be immediately posted back to command to demonstrate our total confidence in him'. Dagworthy was 'stunned', and it took him 'some time to compose [himself]'. Commodore Griffiths – who later became a rear admiral – was supportive and understanding of what Dagworthy was going through.

Dagworthy returned to HMAS *Coonawarra* and was reunited with his remaining crew. He told them what had been recommended and some of them said that they wanted to return to sea with him.

One of the crew, Doug Cousins, said, 'I'm a bushy. Where I come from, when you're thrown off a horse, the first thing you do is climb back on that horse again. What they're doing is the right thing, and I want to come back to sea with you.'

A few weeks later, Lieutenant Dagworthy was back at sea in command of the patrol boat HMAS *Bombard* operating out of Sydney. He had been in command of *Bombard* before he was posted to *Arrow* and was relieved to be returning to a ship he knew well. The first patrol he conducted on *Bombard* was down to Bass Strait for oil-rig surveillance and support. Bass Strait has some of the roughest waters in Australia. They sailed down to the strait and hit bad weather.

'I really struggled,' says Dagworthy. He nevertheless remained in command of *Bombard* for the normal duration of his posting.

Able Seaman Robert McLeod was awarded the Australian Bravery Medal in September 1977 for assisting fellow crew members of HMAS *Arrow* who were injured during the disaster.

CHAPTER 21

SHOAL BAY COMMUNICATIONS STATION

Before Tracy, a new communications station at Shoal Bay was being built on salt flats in a remote area about a fifteen-minute drive from *Coonawarra*. The main gate and guardhouse were up, but the radio station itself was a shell and a long way from being operational. With the original radio station at *Coonawarra* destroyed, it was decided to set up a small part of the new station with radio sets and commence operations as soon as possible. The pressure was on to get Shoal Bay up and running.

With his house uninhabitable and his wife and children evacuated south, civilian technician Michael Pilkington was living and working at Shoal Bay.

He says, 'As you come into Shoal Bay, on the left-hand side of the gate there's a vertical array, and on the right-hand side there's another couple of antennas. They all feed into the radio sets if you've got the right buttons.'

The antennas had been destroyed in the cyclone. The CEO of the company that built them came up to Darwin to inspect the damage. He had built the same type of antennas in Antarctica where they were subjected to severe winds. The CEO and Pilkington drove out to the four antennas. While walking around, the CEO picked up a bolt – about five centimetres in a diameter – and studied it.

The CEO looked at Pilkington and said, 'That's sheared. See there? You had one hell of a high wind.'

'He couldn't believe it,' says Pilkington. There was a weather station out front near the guard building, he adds, that 'was completely blown away. It was gone, so they don't know what speed hit Shoal Bay.'

Pilkington, like other naval personnel and civilian technicians, also had to deal with wildlife, such as the water buffalo that got its horns tied up in the guy wires of an antenna.

One of the men asked, 'What are we going to do?'

Pilkington replied, 'Well just go and pat it on the head and tell it to piss off.'

The chap said, 'Well, away you go.'

Pilkington responded, 'No, no. It's your turn.'

The chap started walking toward the buffalo. It pawed the ground and snorted, 'Whoof, whoof!' The antenna wobbled.

'Poor bloke got it free and off it goes,' remembers Pilkington. 'Oh, good fun, that.'

Another time around dusk, one of the men was driving back to Shoal Bay from Darwin. He turned off the highway to find a buffalo standing on the road right in front of him. He stopped. The buffalo took a long look at him, pooped all over the front of the car and walked away.

Pilkington and one of the other techs decided to go crabbing for a break. His mate asked, 'Well, how are we supposed to get the bait?'

'We'll see the cook,' Pilkington replied. The cook at Shoal Bay gave them rump steak in exchange for a share of their catch.

The steak worked a treat and the two men returned triumphant from their crabbing trip. Their boss found out about the jaunt and asked how many crabs were caught.

About five. The cook wanted two. The boss wanted one, and Pilkington and his mate were left with one each.

'We kept a lot of people happy anyhow,' remembers Pilkington. 'I think we got a big feed of crab every now and then. It was a bloody good job. And we kept everything going.'

* * *

When all the classified material at the old radio station had been destroyed, I was reassigned to sentry duty at Shoal Bay. According to Commander Stan Hall's report of proceedings, 'the Shoal Bay Communications Station became operational on 3 January using ad hoc facilities', which would 'not adversely affect the installation of equipment in operational areas'. Lieutenant Beckwith was made officer in charge of Shoal Bay, and, as the ROPs notes, '[a] good deal of the installation work was carried out by Wrans under the supervision of the project engineer and limited technical staff'.

As each radio set was installed and connected, a couple more radio operators were reassigned from clean-up duties to communications duties. At first, my duties were in security, in rotation with Ruth Turnbull and a few other Wrans. We kept a solo watch in the guardhouse by the only entry gate, accompanied by King, a part-German Shepherd stray the sailors had adopted some time back. He was a sweet-natured dog and good company. Sailors often took him to the beach for a treat in their time off and he was well loved.

King and I would open and close the gate for authorised personnel and do regular security checks of the fence line with a rifle I still didn't know how to use. Whenever cars that were obviously not Navy rolled up, I'd come out of the guardhouse and approach the gate with the rifle in one hand and holding King by the collar with the other, hoping we looked intimidating.

I had coffee with my divisional officer, an Army captain, in the mess at Shoal Bay one evening. I had often babysat his daughters before the cyclone. They'd lost just about everything, and his family had since been evacuated. He had saved up and bought his wife a pair of diamond earrings for Christmas, and she'd never even gotten to see them. They were gone along with everything else. Previously, during a posting overseas, they'd put most of their belongings in storage and lost the lot in the 1974 Brisbane floods. He said their photographs were the hardest things to have lost. Wedding photos and baby photos can't be replicated.

Ruth Turnbull absolutely loved sentry duty at Shoal Bay, but she was not at all keen on the wildlife there. One night, coming off sentry duty, she was walking to the bus with a few others when she spotted what looked like a stick in front her.

'One of the guys said, "Oh, Ruth!", and I knew it wasn't a stick,' she says. It was a King Brown snake, and 'they reckon I didn't touch the ground for about a hundred feet', she adds.

Turnbull also remembered a civilian technician who opened up the cable drains in the building one day and found 'all these baby King Browns in there'. Turnbull was petrified.

When Shoal Bay was being built in 1974, Mick Hamilton was involved in wiring up the coaxial cables, moonlighting at night for Plessey. He was warned to be careful of snakes, which were nicknamed 'Travelling Coax Cables'.

He recalls, 'I had to go down to the beam-forming hut and terminate all these cables,' and whenever he or the other techs unlocked the door of the hut and gingerly opened it, they'd 'see a snake inside the door'.

CHAPTER 22

THE DEATH TOLL

In 2021, Keith Chay wrote to me, 'We can look back now and realise we are part of history. The loss of lives counted at seventy-one would have been expected to be far greater considering the city was virtually destroyed.'

During the clean-up, I sat chatting with a serviceman on our coffee break. He told me that a few days earlier, he and his squad, while on clean-up duties in a Darwin suburb, came across the bodies of a family of five who had been killed in the cyclone. He was bewildered that they had still not been added to the death toll. Richard Usher remembers hearing about this too. Around 6 or 7 January, he was working with one of the daily shore parties around the Nightcliff and Rapid Creek areas with sailors from HMAS *Brisbane*. As Usher recalls, *Melbourne* sailors were cleaning up in the Casuarina area. Usher soon heard the rumour 'about a *Melbourne* team finding a family of five'.

Usher has always remembered that story and has mentioned it many times. He believes there's 'an element of truth to it' and wonders why the family was not added to the official death toll. He recalls talk of the government not wanting a bigger death toll to be revealed, as that 'would have frightened people from returning'.

Although officials were adamant the death toll was correct, there seem to be too many anomalies like the one above.

Certainly, the toll is different whenever the subject is brought up. The possible total of seventeen lost at sea, presumed dead, is rarely even mentioned. I have sometimes wondered if perhaps the toll only included people whose bodies were officially identified. Meanwhile, we had real concerns for the many civilians we knew. We didn't know whether they had been safely evacuated or killed, and their names not added to the toll.

We had no TV, limited radios and no newspapers for weeks. Rumours were rife, and we had no way of investigating or clarifying the stories we heard.

Even after all this time, there remains confusion in my mind about the death toll. The opening paragraph on the front-page article of the *Royal Australian Navy News*, on 17 January 1975, reads, 'Cyclone Tracy, which struck Darwin and stunned the nation on Xmas Day, left forty-nine officially dead'.

According to the Fact Sheet 176, 'Cyclone Tracy', held in the National Archives of Australia, published on 22 May 2003, forty-nine people died in the city and sixteen were lost at sea.

In his 1997 book *Winds of Fury*, Keith Cole lists the names of fifty identified fatalities from the cyclone, and one unidentified body. In addition, he writes, 'fifteen persons are missing at sea and are presumed dead'.

I've read numerous reports and none of them agree on the death toll. I just don't know what to believe, and I'm not the only one. AB Cook Michael Daly of HMAS *Advance* was told by a Navy diver working in Darwin Harbour that when 'they started finding the wrecks, they were often just following the sharks'. Daly says, 'God knows how many people actually died on boats out there because most of them were on charter' for Christmas Eve parties.

Then there were 'hippies' at Mindil Beach, which was packed, says Daly. 'Tents, caravans, all sorts of things just got blown away, so many people must have got blown away as well – all the transients that were up there at the time. There were thousands of them. Who knows how many of them were killed?' he adds. Though Daly admits he can't verify the rumour, 'There was very strong word in Darwin in '75' that a 'couple of freezer trucks full of bodies were taken south'. This rumour, which persisted for many years, has been refuted by authorities since it was first aired. It's just an example of the confusion, the misinformation and lack of reliable information that confronted us daily.

According to Daly, a reporter who was in Darwin at the time had 'a go at' the administrator of the NT as recently as 2020, saying 'the death toll should be amended', as he 'has a lot of anecdotal evidence there were a lot more killed'. Daly himself witnessed at least thirteen dead bodies in one funeral parlour. Daly confirms Usher's 'understanding' that 'the authorities didn't want to dig too deep or say anything because they wanted people to come back, and if the real estimated death toll had been published it may have been the end of anyone wanting to live in Darwin'.

* * *

In the middle of the crisis, staff at the hospital were overwhelmed and put a priority on providing medical assistance before paperwork. They often had limited time to take note of the names of the injured let alone their addresses, dates of birth and next-of-kin. People brought bodies to the hospital and to the police station and left them there without giving their own names or the names of the deceased.

The police set up a temporary mortuary at the Casuarina Post Office and by the evening of Christmas Day, twenty-three bodies lay in state there. Only ten had any forms of identification on them. All the bodies that came through the temporary mortuary were photographed and police took note of anything that could contribute to their identification such as clothing, scars, deformities, jewellery or paperwork. Many of the children were not identified for several days.

There is no doubt the local police and medical personnel did a phenomenal job in desperately tragic and chaotic conditions. Most of them would have lost their homes but despite their own losses, they fronted up for duty day after day. Their work was shattering.

CHAPTER 23

ENTERTAINMENT

During the clean-up, we managed to get some desperately needed entertainment and light relief. The drive-in opened for business well before the screen had been completely repaired. We watched one movie with five panels missing from the screen. No-one ever sat in cars at Darwin's drive-in – it was always too hot. Instead, we sat on 'click-click' chairs in front of the vehicles, often with eskies of beer or soft drinks.

One night, I saw a married sailor and a Wran lounging back on their chairs, in front of their sandblasted vehicle. They were wearing daggy clothes, drinking champagne and watching a movie on the screen with missing panels. They reminded me of the *M★A★S★H* crew – people reacting to abnormal situations in bizarre ways. It's a good coping mechanism.

Colour TV started transmissions in early January 1975, with the first broadcast being the Ashes Test match between England and Australia in Sydney. Barry Chester's business had a few colour TV sets for training purposes, and they had survived the cyclone. Chester loaned two of them to his brother-in-law, Lieutenant Richard Usher, for use on HMAS *Brisbane*. 'Admittedly,' says Chester, 'not all programs were in colour, but the sports were, and that went over well.'

Usher remembers walking into the wardroom where the officers stood around the TV, watching the Test match in colour. Sometimes the colour would fade to black and white, but then return.

Although Sundays were a rest day, Usher says sailors could go ashore to assist family and friends, if they wanted. Usher would help out his brother-in-law as much as he could but remembers that Sunday nights were fun times in the wardroom.

'We would have a special meal and then watch a movie,' he says. In gratitude for the loan of the TVs, Usher invited Chester and his colleague Joe Johns (ex USAF) to join them in the wardroom on Sunday evenings. For Chester and Johns, those evenings in the hospitable *Brisbane*'s wardroom with excellent food, good company and civilised surroundings were a welcome change after weeks of hard work and the monotony of endless stews from the public kitchen.

Watching the Ashes on TV in colour wasn't the only entertainment enjoyed by the crew on HMAS *Brisbane*. When the ship finally came alongside the wharf after being anchored in Darwin Harbour for weeks, the crew was able to 'do clean-up during the day and come back to the ship and get showered up', remembers Simon Gourlay. They were then able to 'step ashore, walk up to the city, or whatever', he says. On one occasion, Gourlay and about eight crew members hired a Mini Moke and drove out to Berry Springs. They also went to the drive-in 'with a heap of guys in this Mini Moke' to see *Alvin Rides Again*. The 'police turned a blind eye' at the overcrowded Mini Moke, according to Gourlay, knowing the men 'were there doing a great job'.

The Navy provided sailors with opportunities for entertainment and sport. 'We worked hard, and they made sure that we weren't going to be bored and do stupid things,' recalls Gourlay. This meant keeping the men occupied during the day and in the evenings.

Gourlay also remembers talk of a lot of looters, which 'the police tried to stamp out'. He witnessed one guy taking a set of eight balls from a badly damaged pool parlour.

* * *

Though children and most women were evacuated south after the cyclone, we Wrans were asked to volunteer to stay and help with the clean-up.

'Being young and silly, we volunteered,' says Lyn Alderton. 'However, I have since learned how few of us there were – I thought there were at least ninety Wrans left, but I believe the figure was more like forty or at the most fifty.' Alderton was one of five Wrans and five married sailors who were sent to Shoal Bay to assist with setting up communications. They were grateful to be working and sleeping in air-conditioning, 'but were isolated and in desperate need of entertainment as were all other crews in Darwin'.

After a few weeks, they heard the Darwin cinema was up and running for the clean-up and construction crews.

'We didn't care one jot which film was showing,' says Lyn. 'Just to get out of the station and into town to see some kind of civilisation was enough for us.'

They found out the evening session time and Lyn, another Wran and two or three off-duty sailors scrambled into their distinctive orange Department of Public Works truck and 'went to the big smoke'. Arriving a bit late, they 'stumbled into the darkened cinema just in time for the short feature'. At intermission, when they were mingling in the foyer and stretching their legs, however, the sailors insisted they had to leave, 'Now!' Alderton and the other Wran were disappointed, but to keep the peace they piled into the truck and headed back to the bush. On the way back, Alderton asked the guys what the rush was to get out of the cinema. The sailors said that the cinema was packed with men on the reconstruction crews – only men – and the movie being shown was *The Erotic Adventures of Zorro.* Discretion being the better part of valour, the sailors decided that a hasty retreat was wise for the only two women present.

* * *

Rolf Harris also came to Darwin during the clean-up to offer some light entertainment for the clean-up crews. He performed on HMAS *Melbourne*, then put on a show at the outdoor amphitheatre in the Botanic Gardens, which I thoroughly enjoyed.

Many years later, we were shocked when Harris was charged with several counts of paedophilia. He was found guilty in court and imprisoned. The memory of his concert is forever sullied by this knowledge.

On 18 January, the withdrawal of the fleet commenced. The weekend was kept clear for the crews of remaining ships to participate in organised sports, bus tours and safaris into the Darwin hinterland.

CHAPTER 24

PROGRESS

Operation Navy Help Darwin remains historically the greatest RAN effort in the relief of suffering due to a weather event.

There were over a thousand sailors working in Darwin. If the days that each man worked were all added together, then the ships' companies performed 17,979 working days on clean-up, salvage, repairs and installations in Darwin up to the departure of the last ship on 29 January 1975. The crews cleared 1593 blocks and the surrounds of residential blocks, schools, hotels, commercial and government buildings, and set to rights the Parap and Nightcliff swimming pools. Repairs were also made where possible. Reroofing and weatherproofing had been carried out on houses in the Rapid Creek area and on priority buildings in Darwin, including the YWCA, some hotels and motels, and the brewery.

The electrical task skill groups assisted the PMG and Electricity Authority in house rewiring, general repairs to exchange and terminal equipment, power-line repairs, and generator installation in large complexes. They even worked on places outside Darwin such as the leprosarium, quarantine station and Gunn Point Prison Farm. Their efforts to repair and operationalise the air-conditioning at Darwin Hospital and various hotels and motels were particularly appreciated.

The miscellaneous tasks performed by naval personnel were many and surprisingly varied. For instance, they helped to save rare trees in the Botanic Gardens; unloaded merchant ships; performed watchkeeping and firefighting duties at Darwin Fire Station; and designed and constructed a water tower at the Mindil Beach caravan park, thus enabling the park to be used for temporary accommodation.

The clean-up of residential properties made the biggest and most obvious impression and impact on the residents of Darwin. Housing and general debris was cut up and removed for possible future use in reconstruction, but most of the debris was carted away for dumping. The care and custody of personal property found on residential blocks being cleared by Navy crews was handled as sensitively as possible, though precious photo albums and books were often unsalvageable and regretfully dumped.

Team leaders brought back other valuables that survived the wind and water damage to the shore headquarters, where they were logged into a register of valuables, then placed in a locked wagon. The items were eventually transferred to cell number eight at Darwin police headquarters. On 18 January, territory police agreed to assume responsibility for the valuables and any further recovered items were handed by the team leaders to police headquarters.

The large naval relief force of physically fit and well-organised personnel not only greatly contributed to the clean-up of Darwin, but also didn't place any pressure on the city's resources in terms of accommodation, food or medical requirements.

The fleet helicopters were also heavily tasked during the clean-up operations. They flew 313.3 hours, lifting 7832 personnel and 110,906 kilograms (244,518 pounds) of stores and equipment. On 8 January, for instance, the fleet helicopters were tasked with lifting material and equipment to the roof of tall buildings such as the Territorian Hotel and the Travelodge.

This invaluable aid to repairs contributed to the town's ability to provide accommodation to displaced residents and civilian work parties.

In addition, the helicopter crews carried out aerial photography and body searches and were heavily involved in surveys undertaken by personnel from the Department of Housing and Construction.

* * *

The assistance the fleet's personnel gave HMAS *Coonawarra* was much appreciated by all of us who were based there.

'Dressed in the remnants of their service and civilian clothing,' reported NOCNA Eric Johnston, '[local personnel] began living in the most uncomfortable circumstances.' We cleaned up and carried out what repairs we could on our single accommodation and married quarters. Fleet personnel gave more skilled assistance with repairs and then shipwright teams were posted in on a rotation basis to make permanent repairs. Johnston's report continues: 'I know of no more heartening sight than the new roofs going on to shells of buildings, ensuring personnel a dry night's sleep and an opportunity to dry and air their clothing, bedding, and store personal possessions which had been recovered.'

Tony Vine of HMAS *Supply*, along with a dozen technical sailors, was assigned to replace the roofs of the new Wrans quarters and the senior Wrans block at HMAS *Coonawarra*. Vine says:

> Our party was under the charge of *Supply*'s chief shipwright and our task was to make the relatively new but damaged Wrans quarters watertight. There were two buildings, the larger Junior Wrans block and a smaller block for the Senior Rate Wrans. Both were multi-storey, and both had flat roofs consisting

> of long sections of overlapping 'W' section guttering. On both buildings, between a third to a half of the roof had been ripped off, some sections landing hundreds of metres away.

Along with the salvageable roofing on the ground, the chief shipwright estimated that there was enough good roof remaining on the smaller building to repair the larger block's roof. The team then stripped the roof off the smaller block and carried it over to the larger one. This was in the days before workplace health and safety, and the men worked shirtless, three floors up, without safety harnesses and hard hats. Nevertheless, they managed, without incident, to make the junior Wrans block watertight within a couple of days. In the meantime, 'a truckload of bright, newly milled, one-metre-wide aluminium sheets' arrived to repair the smaller block. They set to work under the guidance of the chief shipwright who, says Vine, 'was determined that no wind was going to blow the new roof off'. The shipwrights constructed a slightly gabled roof and fixed it to the structure with RAMSET guns.

The dangers of working so high up became apparent when a young able seaman, carrying a sheet of aluminium along the roof, was hit by a gust of wind that turned the sheet into a wing.

'He was off and flying,' says Vine, 'and was unable to let go.' A quick-thinking team member tackled the young sailor back onto the roof and the sheet of aluminium sailed off across *Coonawarra*.

'We also found to our discomfort that the reflective qualities of the sheets were such that we were all sunburnt under our armpits, and worse still up our shorts and in our crotches by the reflected sun as we laid the sheets,' says Vine. 'Yes, we were sunburnt where the sun does not normally shine!'

* * *

When General Stretton had arrived in Darwin on 26 December 1974, he appointed NOCNA chairman of the port committee, giving him operational control of the port and its approaches. NOCNA determined that the survey vessel HMAS *Flinders* be employed in fixing the position of navigational aids and locating harbour obstructions.

Clearance Diving Team One, which had arrived in Darwin on 26 December, carried out extensive diving operations around the wharves. Their main aim was to remove the sunken patrol boat HMAS *Arrow*. After that, wharf repairs could be undertaken.

The CDT successfully refloated *Arrow* on 13 January 1975 by attaching pontoons to her hull. Tugs pulled her clear of the wharf at high tide. She was towed underwater to shallow water at Frances Bay where she was surveyed and written off. It was gut-wrenching to see her lying forlorn on the mud flats, severely damaged, beyond repair, abandoned. *Arrow* eventually broke up where she lay.

Eric Johnston praised the diving team for their tireless efforts. 'They had undertaken general harbour search and surveys; they had cleared the wharves and immediate approach areas and they had salvaged and raised the wreck of HMAS *Arrow*.' He goes on to say he 'lost his temper' only once 'during the whole clear-up operation', when 'a well-known television commentator, encountering the clearance divers during one of their rest periods, asked them why they were bludging'. He grimly added, 'This man does not know how close he came to joining the other debris in the harbour.'

* * *

Departing Operation Navy Help Darwin wasn't easy for some. Petty Officer Medic David John was advised he would be returning south after only five days, with some more personnel who'd been seconded from other ships and establishments.

They were taken to the RAAF Base at 4 am for a flight back to Sydney later that morning. 'As with all things in the military, it then became a game of "hurry and wait",' he says. Their C130 Hercules didn't arrive until 4 pm, and then had to be unloaded, making it 5:30 pm before it took off for Sydney.

In the spirit of the Navy being self-reliant and not imposing on civilian resources during the clean-up, only military aircraft were to be used for military personnel. While waiting for a Hercules, John and his companions watched with frustration while commercial aircraft flew back to Sydney, Brisbane and Melbourne without any passengers all day.

They arrived at RAAF Richmond at around midnight and were then transferred by bus to HMAS *Kuttabul* in Sydney where they were left to their own devices. John caught a taxi to his married unit at Coogee and arrived there at around 3:30 in the morning. He had not been able to phone his wife from Darwin to let her know he was coming home, and knocked on the door for several minutes before his wife warily asked, 'Who's there?'

She looked through the peephole, but John had grown a beard during his deployment, and she didn't recognise him. She refused to unlock the door until she was sure the man on the other side was indeed her husband and then finally opened the door to welcome him home.

* * *

By 8 January, 496 civilian houses and 110 Navy married quarters houses had been cleared. Fourteen houses had been reroofed, and ten others had temporary roof repairs completed. The work teams were now in a set routine.

Some of the ships were berthed at the wharf but most were anchored in the harbour. Clean-up crews were transferred to and from shore by landing craft, tugs or even helicopter. HMAS *Hobart*'s relief parties were transported ashore in

Darwin Harbour Board's tug, *Goyder*, at 6:15 am every day, then they proceeded to headquarters by bus. There, the teams collected tools and equipment and were transported by bus to their assigned housing blocks.

Hobart's chefs prepared lunches and iced limers. A duty victualling officer went ashore with the lunches at 10:30 am daily and travelled in the delivery van to assist the driver in locating the teams.

Clean-up crews from other ships collected their lunches to take with them when going ashore every morning. Work ceased at around 3:30 pm when buses collected the relief teams and returned them to headquarters. The teams returned their equipment and reported their progress, before reboarding buses for the trip back to the wharf. *Hobart*'s crew were returned to their ship by the tug *Goyder*.

In addition to cleaning up residential neighbourhoods in Nightcliff, teams from *Hobart* were tasked with cleaning out two of Darwin's public swimming pools. The debris was cleared, the grassed areas were mown and the rebuilding began. Shipwrights repaired the roofs and engineers repaired the filtration units. Seamen scrubbed the pools with disinfectant and detergent. Within a week, the pools were available for use.

* * *

On the evening of 8 January, NOCNA decided to change the focus of Navy work teams to assist with repairs to houses of residents who had remained in Darwin. That evening, an announcement was made over the radio. Residents from any area who required assistance in cleaning up and repairing their houses were invited to contact shore headquarters so a team might be allocated to them, and a time established for their arrival. The only stipulation was that one resident must be present to work with the team. In the next few days,

580 requests were received, and the list was temporarily closed to prevent a backlog.

Every day, the *NT News* listed the homes due for team assistance the following day, and the scheme was an outstanding success. Close liaisons developed with the locals, and many lasting friendships were commenced. Concurrent with these private requests, general clearance in the Nightcliff area continued until 17 January.

Sunday 12 January was declared a rest day for the fleet, but even after eleven continuous and arduous days of work, 414 volunteers continued to work on homeowners' blocks or to assist friends, relatives or people they'd worked with during the week. On that day, they cleared 126 blocks, which was double the daily average.

When HMAS *Supply*'s Tony Vine finished replacing the roofs on the Wrans quarters at *Coonawarra*, he put in one last day at Nightcliff on a volunteer Sunday work party, before he found himself again watchkeeping on board *Supply*. Work on Sundays was voluntary and there was never a shortage of volunteers. He says:

> On one of these days, a young sub-lieutenant pushed a work party too far. The parties were issued with water, fruit and sandwiches, which by lunchtime were generally not very palatable. On this particular Sunday, the owner of the property the team were working on managed to put on a small barbecue for them. Unfortunately, young 'sir' arrived and ordered them not to eat the food and said that they had to eat their sandwiches. A radio operator, who was paying off when we got back, said something along the lines of: 'Well I suppose it better be me', stood up and dropped the subbie.

The landing craft *Betano* and *Balikpapan* had arrived in Darwin just after sunset on 4 January 1975. On 6 January, their rest day,

Balikpapan's company worked on clean-up duties in Nightcliff. The next day, they returned to sea. *Betano*'s main job, according to the commanding officer, Lieutenant Graeme Rogers, was to collect and return *Melbourne*'s and *Supply*'s working parties, as well as 'daily sullage runs, additional personnel runs, stores delivery and delivery of fresh water'. The work commenced at 5:45 am and finished at 10:30 pm. The arrival of the three LCHs (landing craft heavy) – HMAS *Brunei*, *Tarakan* and *Wewak* – in Darwin provided *Betano*, once their cargo had been offloaded, with some respite 'to carry out necessary maintenance as well as enjoy a brief stand-down period'. *Betano* was out of action for two days, resuming general duties again on Thursday 16 January and remained in Darwin for another week, providing general support to the task force units in Darwin Harbour.

CHAPTER 25

SURVIVORS' LEAVE

The Navy task force performed sterling work on clean-up and repair duties. The presence of so many naval personnel also enabled those of us stationed in Darwin to have some desperately needed time off.

I worked sentry duties at Shoal Bay until I went on leave in mid-January. It was officially called 'compassionate leave', but we knew it as 'survivors' leave'.

Naval personnel from HMAS *Coonawarra*, *Melville* and the patrol boats were to be flown out in stages for ten days' leave. We needed a break. While many sailors and Wrans had been on leave and returned to work by 14 January, I still had not been scheduled to go. That day, while on sentry duty at Shoal Bay, Radio Supervisor Special Cooper delivered a letter to me from my mother. Her letter was full of loving concern and suddenly, I began to cry. I sat still and silent, unable to speak, powerless to stop the flow of tears. I'd reached the limits of my capacity to cope.

RSS Cooper was wonderful and organised for me to be relieved from duty and to see the naval doctor. I sat in the doctor's surgery feeling like a fool, still unable to speak, while he arranged for me to leave Darwin on the next flight out. Bernie Kloss later sought me out. He held me close while I cried and gave me what quiet comfort he could. We all had so much pain.

I flew out to Brisbane on 15 January and stayed overnight there. It seemed like a different world. Everything was so normal and untouched by disaster. Rockhampton, too, was so normal that I couldn't fit in. Despite the kindness of my family, I felt out of place. I'd somehow become used to living and working in a disaster area and normal life felt odd.

Trying on new clothes and listening to the shop assistants complain about the increase in food prices seemed unreal.

My mother and I visited an elderly relative who mildly scolded me for not wearing my uniform home. I told her that I didn't have one. She became indignant, saying something like, 'You're in the Navy – of course you've got a uniform!' People didn't realise just how much we'd all lost. It wasn't just a big storm.

* * *

Able Seaman Michael Daly of HMAS *Advance* had a far from restful time on survivors' leave. He and AB Ray Minette decided to drive to Melbourne in Ray's car. With only two weeks' leave, that was a huge undertaking.

'We didn't muck around,' says Daly. 'We got to Alice Springs in, I think, two days.' They drove down the Stuart Highway and crossed over to South Australia where the bitumen ended. The dirt road was rough, heavily corrugated and damaged by road-train traffic.

Minette was, according to Daly, driving a little bit too fast for the conditions when they came across the body of a dead kangaroo in the middle of the road. Minette tried to steer around the carcass. The tyres skidded on the bull dust, and the car rolled five or six times. They didn't have seatbelts on, and Daly reckons he 'went out the window twice'. He remembers thinking, *We survived Tracy and we're going to get killed now in a car accident in the middle of nowhere.* The car was wrecked and their belongings had flown everywhere. When a road train finally came along, the truckie who took pity on them put some air in the tyres, tied the car to the back of his truck and towed it to the nearest police station. Then the two sailors went to the bar and got drunk, before the driver took them back to Alice Springs. They ended up having to 'spring for an airplane ticket to Melbourne'.

Once home in East Bentleigh, Daly bought an HR Holden. Although he didn't have a driver's licence, Daly decided to drive the car back to Darwin. All the way to Adelaide, he drove sedately hoping 'the boys in blue' wouldn't pull him over. In Adelaide, he put the car on The Ghan, not risking driving on the dirt road again. Few of us had so much trouble surviving survivors' leave.

* * *

At the end of January 1975, HMAS *Stuart* departed from Darwin, towing the patrol boat HMAS *Attack* to Cairns for repairs and a refit. Minimum crew members stayed on *Attack* for the voyage. The rest of the crew were given ten days' survivors' leave. The crew were given plane tickets, and Bernie Kloss was set to fly down to South Australia, where his parents lived. His flight was scheduled to leave Darwin at 2 am.

'Of course,' says Kloss, he and the other sailor travelling with him 'had a few cool drinks before two o'clock in the morning while we were waiting to get on the plane'.

They were tired when they arrived in Adelaide, so they booked into a hotel, went straight to bed and slept most of the day. The stress and sleep deprivation of the past few weeks caught up with most of us sooner or later. The next day, Kloss was invited to visit an old mate from primary school. They hadn't seen each other for a few years. Kloss says his friend's mother cooked a chocolate cake for the occasion – Kloss's twenty-first birthday – and took a photograph of him cutting the cake. Having made it to twenty-one, Kloss thought the rest of his life 'could be pretty interesting'.

He hired a car and went up to his hometown in South Australia to see his mum and dad. Communications were pretty poor in Darwin after the cyclone, so his family hadn't known how their son was faring until he finally got through on the phone after about ten days or so. He spent a few days with his parents and two brothers before flying up to Cairns to work on the refit and rebuild of HMAS *Attack*.

Like most of us, Bernie found it hard to talk to people about our experiences during Cyclone Tracy and the conditions we had worked under. People had seen the footage on TV and the photos in the newspapers, but that's very different from having actually been there. Nobody wanted to distress family and friends by going into too much detail about our losses and the conditions in which we were living and working.

* * *

Two Wrans, Lyn Alderton and Kaye Marshall, were some of the last in our group to depart on survivors' leave. Alderton had been living and working at Shoal Bay Communications Station. Around the end of January, she was advised she'd be going home on survivors' leave for two weeks 'sometime soon'.

'I packed my meagre belongings in a small bag and looked forward to a break from living with only ten or so other people at the new station down the track,' says Alderton. On 6 February, she went back to *Coonawarra* to pick up her travel papers and leave pass. She was given instructions to arrive at Darwin Airport in plenty of time to pass through the exit procedure, which would require more paperwork at the check-in desk. Marshall, a good friend, joined her at the base as they were travelling on the same flight to Brisbane. Marshall would then travel on to Melbourne.

The two Wrans fronted up to an official, probably a federal police officer, at the check-in counter at the airport. Alderton and Marshall handed over their tickets and the officer said they would each have to complete a departure questionnaire to receive the re-entry permits they would need on their return to Darwin.

The last question was a surprise: 'Do you intend to return to Darwin?' or words to that effect. Alderton thought, *Is this guy not with it, or what? Because if we don't return by the date when our leave pass expires, one of his colleagues is going to have to come looking for us!*

When the pair arrived in Brisbane they were 'taken aback by the number of people and the noise'. It was early evening and the airport seemed hectic and chaotic to Alderton. She rang her parents from a public phone at the airport because she hadn't been able to contact them from Darwin to let them know she was coming home on leave. She had to wait for an hour or so until they collected her.

Marshall also had to wait for her connecting flight to Melbourne, so they sat side-by-side in a row of seats and tried to adjust to the 'normality' of a busy city airport. They chatted for a while and then Marshall had to board her flight. Alderton recalls:

> That parting was one of the hardest goodbyes I have ever experienced. I *knew* we would be back in Darwin in a few weeks and that I would see her again, but it was difficult to let her go that day. When she left, I felt like an alien from another planet.

* * *

It was during our survivors' leave that many of us started to feel the disconnect between those of us who had endured Cyclone Tracy and those who hadn't. 'What was it like?' was the question asked by family, friends and casual acquaintances. Where to start? We'd have a go at telling people but, with the best will in the world, they couldn't comprehend the enormity of the experience. They all had their own problems, obligations and interests, and soon moved on. This was perfectly understandable.

We couldn't fully explain what 'it' was like to people who hadn't been there. We didn't like to open up with other people who had been there too because, although they would understand, they were suffering as well. We didn't want to add to one another's burdens. The clamming up and closing down began.

PART 4

LIFE GOES ON

'The need for a navy will never go away. History has not finished with us.'

Ben Wilson, *Empire of the Deep: The Rise and Fall of the British Navy*

CHAPTER 26

WINDING UP OPERATION NAVY HELP DARWIN

HMAS *Melbourne*'s commanding officer, Commodore GR Griffiths, praised the 'background plan' used in the Navy's Darwin operation: a plan which had its origin in a naval exercise the previous year.

'We set it up in day one – we are still using it and we will continue to use it,' he said, adding, '[t]he operation went extremely well.' Asked if he could see any room for improvement, Commodore Griffiths said, 'We should look at the stores we take with us. Some equipment, such as chainsaws, were in short supply.'

With the city returning 'to some sense of normality', as Brian Swan of *Melbourne* puts it, the Navy's 'work came to an end'. They'd been there for seventeen days and he had personally travelled over 2736 kilometres in his utility vehicle. 'We hadn't shot anyone, and we'd got the city working again,' he says. Civil tradespeople were pouring into Darwin to take their place, and it was time for the withdrawal of Navy vessels to begin.

The patrol boat HMAS *Attack* had received temporary repairs but was not operational. Major repairs and a refit were required but they could not be performed in Darwin.

While 'the generator was still okay', according to Bernie Kloss, 'there was nothing left of the propellers'. The blades had been ripped off the propellers when the patrol boat, going full stern, had hit the rocks. One of the propeller shafts was also bent, making the boat undrivable.

The destroyer escort HMAS *Stuart* was assigned to tow *Attack* to Cairns, while also providing an escort for the three LCHs, HMAS *Brunei*, *Tarakan* and *Wewak*. *Attack*'s chief engineer had come back and only a minimum crew was required for the voyage, so Kloss took survivors' leave.

No patrol boats had previously been towed for such a long distance – 2000 kilometres – and considerable care was taken to ensure *Attack* could cope with the strain. *Attack* was moved onto the Frances Bay slipway, patched, de-propellered and given self-steering capability.

The little armada departed at 11 am on Friday 17 January 1975. *Attack*, with her commanding officer Lieutenant Paul de Graaf and six sailors, rode easily at the end of the 150-fathom towing hawser. On arrival in Cairns on Wednesday 22 January, the Cairns tug took over from *Stuart* and towed *Attack* safely into harbour.

* * *

After the clean-up, the city of Darwin became 'a great supporter of the Navy,' says David Shackleton. A few days before their departure, he recalls 'a sailor from *Melbourne* putting "Navy came to Darwin" bumper stickers on cars', and being told by a local, 'you don't need to do that, mate. We will always remember that it was the Navy that helped us when everybody else left town.' Shackleton says they were 'chuffed' that they'd 'made a difference'.

Meanwhile, 'the world was moving on,' he says, 'and more relief effort for Darwin was arriving from the south. The emergency relief provided by several RAN ships had been effective in getting things started, but a long period of rebuilding the city was needed.'

On 18 January, *Melbourne* left Darwin for Sydney to prepare to lead the RAN task group to take part in the naval exercise RIMPAC-75.

* * *

HMAS *Supply* was a Tide class replenishment oiler whose role was to refuel other naval vessels at sea. Replenishments at sea (RAS) are carried out almost anywhere at sea, day and night, and in a wide range of weather conditions while ships are underway. Crewman Tony Vine says:

> Late in January, on the day I was to be promoted to leading seaman, I was detailed off by the regulating chief stoker to become the petty officer's messman. The chief stoker had a distinct dislike for apprentices, and he insisted that I would do messman for a minimum two months regardless of my rank or the impact on my watchkeeping progression. I immediately withdrew my request to be promoted and submitted a new request to be discharged. When the ship's engineer officer found out why I was requesting a discharge, he exploded, and the chief stoker was told to change his watchbill. My relationship with the chief did not improve, however, I was promoted on time, and I continued my career progression down in the engineering spaces.

Despite the united effort to provide disaster relief, the normal stresses and problems of life in the services frequently arose like this.

While in Darwin, '*Supply* was kept busy,' says Tony Vine. Nine vessels of the task group came alongside to refuel while the ship was at anchor in the harbour. Vine remembers having to shift anchorage one night when the anchors began to drag due to a high tidal flow coupled with a strong wind. On 18 January, *Supply* proceeded out to sea and fuelled *Melbourne* prior to its departure for Sydney.

After *Melbourne* departed, FOCAF (Flag Officer Commanding Australian Fleet) briefly shifted to *Supply*, making it the 'flagship of the Australian Fleet'. Vine says:

> Three days later we went alongside Stokes Hill Wharf to top up the diesel tanks in the Navy fuel farm using *Betano* as a tug. Unfortunately, the strain on our hawser proved too much for *Betano* and one of her bollards was torn from the deck. Amazingly, nobody was injured, but a lot of people slipped away to change their underpants.

Supply finally sailed from Darwin on Friday 24 January, with *Vendetta* in company. The crew thought they were bound for Sydney. The following day in the Gulf of Carpentaria, the fleet oiler refuelled *Vendetta* which then proceeded independently to Sydney. The news was then broken to *Supply*'s crew that they had been tasked to go to the former HMAS *Tarangau*, the Navy base on Manus Island, north of Papua New Guinea, to empty the RAN's stores of fleet furnace oil from the tanks on the island. Papua New Guinea had recently become independent and there were concerns that the new government might seize the valuable fuel as a source of revenue. The decision did not go down well with the crew, however, so commanding officer John Stevens made sure everybody got at least one day ashore at Manus.

Supply finally arrived in Sydney on 7 February 1975, where the crew was welcomed back by family and friends waiting at the dock. Although it had been six weeks since they'd left Sydney, Vine says 'it felt like three months'. The crew was granted an additional week of leave for their services.

* * *

Most Navy ships had left Darwin Harbour by 29 January 1975. On 30 January 1975, HMAS *Stalwart*'s commanding officer Captain IM Burnside turned over control of clean-up duties to the officer in charge of 57 RAR, Lieutenant Colonel O'Donnell. The Navy's shore headquarters closed down and Operation Navy Help Darwin was officially completed. However, a small crew of Navy volunteers from *Stalwart* stayed in Darwin to help with the reconstruction of naval buildings and infrastructure.

In his lecture 'Operation Navy Help', NOCNA, Captain Eric Johnston, said that he 'saw very closely what the Navy achieved and the manner in which it was done'. He also noted 'the almost total absence of the dead hand of bureaucracy', which 'allowed tasks to be undertaken without worrying about the correct channels or the eventual cost'.

Some of the decisions he made had unwelcome consequences years later when he was serving in Navy Headquarters, Sydney. A memo from the secretary, Department of Defence, expressed horror at the number of vehicles Johnston had requisitioned in Darwin – 'many of them new and many of them returned in very poor and damaged condition'.

He also received a letter from the Navy's legal advisor informing him that he'd 'been cited by an iron ore mining company as being the principal cause of their liquidation'. This was because Johnston had prioritised wharf repair over the repair of a conveyor belt used to move a large iron ore deposit from the port to cargo ships. Johnston was issued with a bill for $16,900, the market value of the iron ore that he appropriated to build a ramp for the landing craft to use to discharge their loads. Johnston added wryly, 'May I say that iron ore when packed with water and sand makes a most admirable slipway.'

* * *

On 31 January, the destroyer tender HMAS *Stalwart* and the guided-missile destroyer HMAS *Brisbane* had the distinction of being the last RAN ships to leave Darwin.

Stalwart had arrived in Darwin on 2 January and landed 150 personnel, who were organised into clearing and salvage teams. The crew had spent the next twenty-nine days participating in Operation Navy Help Darwin either on shore on clean-up duties, or on board in support roles.

Stalwart was the largest naval vessel wholly designed and built in Australia. She operated as a mobile maintenance facility with engineering, electrical, weapons, shipwright and other workshops on board to enable her to provide maintenance on up to four destroyers or frigates simultaneously. More than 75 per cent of the ship's complement were engaged in repair and maintenance tasks. *Stalwart* also had evaporators capable of producing up to 350 tonnes of condensed fresh water from the sea each day. The ship greatly increased the Navy's self-sufficiency and enabled warships to operate for long periods away from their home ports. Her presence in Darwin was invaluable. Not only did she provide maintenance facilities for the other ships participating in Operation Navy Help Darwin, but her specialist crew members performed skilled tasks ashore. Their list of achievements is impressive. *Stalwart*'s crew restored electrical and communications systems and repaired generators, electric motors, air-conditioners, motor vehicles, chainsaws and a multitude of other machines, engines and equipment. They performed boat repairs; reroofed many private and public buildings; and even made structural repairs to the Darwin Hospital, the Rapid Creek laundry, the St John's ambulance depot and the Red Cross centre. Two Wessex helicopters operated from the ship, facilitating the lifting and shifting of heavy materials.

The crews of all the ships in the task force made significant contributions to the clean-up after the disaster. It's hard to quantify how much they meant to the people of Darwin, and the improvements they made to the devastated city and to the morale of the population.

* * *

HMAS *Brisbane* claimed the honours of being the first to sail from Sydney, the first to arrive in Darwin, and the last to leave. While in Darwin, an average of 160 men were landed daily to assist in the clean-up operation. Apart from clearing debris, they reroofed homes and commercial buildings; installed generators; salvaged sunken craft; provided drivers and heavy plant operators; repaired essential services; and repaired or replaced equipment at commercial sites.

Richard Usher didn't have the opportunity to speak to his wife, Bronwen, from the very early hours of Boxing Day 1974 until mid-February 1975. There were no mobile phones, email or Zoom in those days. Usher had been disappointed not to have his camera but realised he would not have been able to buy film for it in Darwin anyway. Always a priority in the Navy, 'the mail got through', says Usher, and 'the pipe on board "Mail is ready for collection" always put a spring in every man's step'. Usher and his wife wrote to each other almost daily. It was their 'only way of maintaining contact, and a treasured one at that'.

When Usher departed Sydney on HMAS *Brisbane*, Bronwen recalls that the news was full of the plight of Darwin for the next six weeks, and their time apart seemed particularly long. While Usher was away, Bronwen says their young son Peter 'continued to grow and develop', and by the time Usher returned home, Peter 'was his own little man'. He felt 'resentful' of his father and didn't want to share his mother with him. Peter 'often put himself between us when he thought we were too close together'. Bronwen says that situation 'didn't last long, but it brought home to me how, despite their tender years, things do affect children deeply and certainly made me very aware of the situation of parents with children who had survived the cyclone and been evacuated'.

Usher's brother-in-law, Barry Chester, says Darwin was truly sorry to see the Navy go. 'By then,' he says, 'most of us had heard of the amazing organisational efforts that got the ships there so quickly, and we had come to appreciate how important was the ability of the ships to sustain so much effort over several weeks without draining scarce local resources.'

Chester believes that 'the Royal Australian Navy has made it its business to adapt and would not be found wanting if – God forbid – we ever face another national crisis on the scale of Cyclone Tracy'.

Simon Gourlay, who spent his time in Darwin working in the pay office on HMAS *Brisbane* as well as doing clean-up duties, felt they 'could've stayed a bit longer'. The ships were scheduled, however, to return to their normal duties or participate in exercises.

'But I think we felt that we hadn't finished what we set out to do,' says Gourlay. 'We'd started it and put a big dent in the clean-up operation itself but could have done a lot more.'

CHAPTER 27

GRATEFUL THANKS

On the withdrawal of the Navy from Darwin, the Naval Board expressed its 'appreciation of the magnificent work by all concerned in Operation Navy Help Darwin' in a signal which was reported in *Navy News*, February 1975:

> Personnel from *Coonawarra*, *Melville*, the Darwin-based Patrol Boats, Task Force 327 have borne the brunt of the task under very trying conditions … The efforts of all concerned have been in the best traditions of the Service and have attracted a great deal of well-deserved recognition and acclaim. Well done.

In his report of proceedings, January–March 1975, Captain Eric Johnston made particular note of the Wrans who 'assisted civilian relief organisations and manned communication centres for the police department, fire brigade and the Department of the Northern Territory'. He also points out their 'performance of duty was such that these departments kept them for much longer than was necessary', and would have continued to use their services if the Navy had not shown 'firmness' in recalling them.

Years later, Johnston remarked on the bond that formed

between RAN members and the citizens of Darwin over January 1975.

'The sympathy and support given by the fleet's sailors and shore-based sailors and Wrans,' he said, 'was matched only by the cheerfulness and togetherness of the citizens of Darwin.' When 'normality returned' to Darwin six months later, Johnston said, he 'looked back with a feeling of pride in being not only a member of the Royal Australian Navy but also a Darwinian'.

Commodore Guy Griffiths, the commanding officer of HMAS *Melbourne,* said, 'The people of Darwin had been grateful for the Navy relief work.' At a reception, Darwin's Mayor, 'Tiger' Brennan, presented a plaque to Rear Admiral DC Wells, fleet commander and commander of the task force group in Darwin. Brennan commended the Navy for its efforts: 'I will not forget you. We owe the Navy the greatest debt of all.'

Despite all this praise for 'the Navy', most of the naval personnel who endured Cyclone Tracy and who worked on the clean-up were never given any credit for our actions. Several well-deserved Naval Board commendations were awarded, but all other naval personnel were ignored.

There are no entries on my service record to indicate I went through Cyclone Tracy and worked on the clean-up. It is noted that I was posted to HMAS *Coonawarra* in Darwin on 19 November 1973; I had compassionate leave from 15 January 1975; and I posted out from HMAS *Coonawarra* on 5 January 1976. There is no mention of Cyclone Tracy. Unless you knew Cyclone Tracy struck Darwin on 24 December 1974, you would not realise I experienced it. There is no mention of my having taken part in Operation Navy Help Darwin or of the work I performed during and after the operation. I don't know for sure, but I suspect this is the same for all Wrans who were based in Darwin at the time.

The lack of recognition for the service of men and women during peacetime kept many of us who worked on the clean-up from joining the RSL and deterred us from marching on ANZAC Day. I attended every ANZAC Day parade as a spectator and held the date dear in my heart but, with no recognition of my service, I didn't feel I had a place in the commemorations. It wasn't until we received the Australian Defence Medal in 2005 that I began to march occasionally.

Sadly though, a lot of Wrans who quietly left the service after Cyclone Tracy didn't qualify for the Australian Defence Medal as they didn't serve for a full four years. Some who left the service because of government policies of the day, such as marriage or pregnancy, are now entitled to the ADM. Those who served in Darwin during Cyclone Tracy and the clean-up, however, have received no recognition for the phenomenal work they did there under such appalling conditions.

The case for Major General Alan Stretton, who headed up the National Disasters Organisation, was the exact opposite of ours. He flew into Darwin a day after Tracy but only spent a couple of days there before heading back to Canberra where he ran operations. I am likely being unfair, but it reminds me of the British Army generals in the First World War who never went to the Western Front, and who directed operations as far from enemy lines as they could get.

Stretton returned to Darwin sometime later after the clean-up work had been done. He toured the city along with Prime Minister Gough Whitlam, receiving accolades for a job well done. Stretton was made an Officer of the Order of Australia (OA), a Commander of the Order of the British Empire (OBE) and was even named Co-Australian of the Year in 1975.

Many people say that we should not expect any recognition. We were doing our jobs, we were doing what we were paid to do. That may be true, but so was Major General Stretton.

Roger Fairbairn, the junior officer who was faced with initiating the RAN's response on Christmas morning at Naval Headquarters in Sydney, remembers Major General Stretton coming in after he'd taken 'responsibility for dealing with the aftermath of Cyclone Tracy'. Fairbairn found him 'overbearing' and thought 'he was full of bullshit'. Fairbairn says, 'He wrote himself up extremely well and didn't mention any of the people who had actually done anything.' A lot of what came out of the newly formed National Disaster Organisation 'was just hot air and illogical thinking', according to Fairbairn. 'They took over, pretty much saying, "We're going to organise everything, and you guys just do what you're told",' he adds.

Petty Officer Mick Hamilton also says, 'there were a lot of heroes that never got a mention.' He cites the example of Able Seaman Taffy Lloyd who 'rescued a family from the XO's house' at HMAS *Coonawarra* and who 'never got a thank you or a mention. Nothing.' He also mentions Ray 'Skull' Morse who dragged Hamilton from the transmitter station to the diesel shed at the height of the cyclone. Hamilton wants these 'blokes to get some sort of recognition because they were heroes. They risked their lives for others.' Yet they received nothing.

Many Wrans and sailors like Mick Hamilton stayed at their posts despite the danger until the buildings disintegrated around them. Naval personnel whether off or on duty did everything they could within their power to protect property and the lives of their fellow servicemen, servicewomen and families.

Bill Ritchie of HMAS *Melbourne* had been supervising sailors for around twenty-nine years at the time of Cyclone Tracy. He found his 'appreciation of sailors' "can-do" attitude was again strengthened by watching the exhausted men returning to their ships each day, having been ashore for a long period in high humidity and heat'. The next morning, he says, 'they would generally front up, without grumbling, and get on with the job. They did this day in and day out until they were replaced.'

Ritchie's 'greatest disappointment' about the whole operation is that no recognition has been 'given to those who participated in Navy Help Darwin. There has been no medal or award given, yet a smaller cyclone later saw awards made.'

* * *

The lack of recognition for our work also puzzles Simon Gourlay of HMAS *Brisbane.* A few years ago, he spoke to a Liberal politician in Darwin, who was intrigued that he worked on relief efforts after two tragedies – the Tasman Bridge and Cyclone Tracy – within such a short timeframe. According to Gourlay, she was pushing to get a Humanitarian Medal through parliament for all those who worked on the clean-up in Darwin after Cyclone Tracy. He doesn't know if the medal idea was ever pursued but thinks 'we deserve something like that'.

* * *

For Tony Vine, the relief operation 'changed his perspective on life and the Navy'.

> It made me realise that what we consider to be our normal lives can literally be blown away overnight, but also the power of the defence force to respond to and assist when natural disasters occur. In recent years, this has become almost routine, and our current generation of people and ships regularly deploy to areas of natural disasters.

Vine says his six-week deployment in 1974–75 'also brought home how many decisions made by those in command can have either an immense positive or negative impact on morale and the overall effectiveness of an operation'. The granting of a second beer issue every evening, which would probably be unacceptable today, played a major part in allowing the work parties to wind down after the day's labour in heat and humidity. Vine also credits his commanding officer, John Stevens, for keeping in touch with his sailors to assess their mood and wellbeing. Another decision that boosted morale, he says, was granting the sailors an extra week of leave to spend with their families after their deployment in Darwin.

'On the other side of the ledger,' Vine says, 'there was much that could have been done better.' For many years, Vine had a poor opinion of the Australian Army. He deplored the Army's failure to release essential stores in Townsville. When the Navy arrived in Darwin, they were told 'the local Army command was intact but had remained in Larrakeyah Barracks after the cyclone, looking after their own'.

'Despite the devastation,' he maintains, 'the Army provided virtually no assistance to the community for the first week even though 30,000 people were homeless.'

To be fair, the Army's presence in Darwin was very small. So small, in fact, that they had to combine with the Navy in order to be competitive in the annual inter-service sporting matches. We often fielded an Army/Navy team against the RAAF in cricket, baseball, basketball, soccer and even squash in Darwin. Army and Navy radio operators worked side-by-side at *Coonawarra* and the soldiers I knew were hard-working, tough and disciplined. Larrakeyah Barracks was totalled by the cyclone and most of their residences, if not all, were destroyed. They carried a heavy load, but their small numbers meant they were hardly visible in the initial relief operations.

Vine considers Cyclone Tracy 'a pivotal point' in his life, both personally and professionally. 'I went up there as a nineteen-year-old, but acting like a sixteen-year-old, and came back as an adult in every sense.' He admits he was naïve when it came to the broader role of the Australian Defence Forces and 'came home with a vastly different perspective'. Vine says he wouldn't have missed 'those amazing and hectic six weeks' for 'all the tea in China'.

> I ended up serving until 2010 in the PNF paying off as a CMDR as senior naval officer Tasmania. The lessons learned and memories of Tracy relief team stay with me today. Like the ABRO, I never suffered fools, but most of the time I behaved. My kellick writer reservist (ex-leading Wran writer) used to warn people that deep inside Commander Vine was WOMTPSM Vine trying to get out.

* * *

The National Emergency Medal, established in 2011, is an award of the Australian honours system given 'for sustained service during a nationally significant emergency; or to other persons who rendered significant service in response to such emergencies'.

There is no doubt that naval personnel involved in Cyclone Tracy and the clean-up fulfil all the eligibility criteria for the awarding of the National Emergency Medal – except for the requirement that the emergency must have occurred in the last three years. That time restriction makes responders to Cyclone Tracy ineligible.

There's a saying that our trials are meant to make us better, not bitter. That has been my own greatest challenge. To focus on the phenomenal dedication to duty displayed by naval personnel in Darwin during Cyclone Tracy and the clean-up. Even though the government and the Navy won't acknowledge our individual contributions, we can be proud of ourselves and our service.

And Darwin is thriving. That's a source of pride and satisfaction, knowing that we played our own little parts in its recovery, as proud members of the WRANS and the RAN. We did a good job.

CHAPTER 28

THE WORK CONTINUED

The fleet had been a reassuring and powerful presence. I can only speak for myself, but their departure left me feeling a bit stranded.

We were fortunate to have exceptionally good leaders. Captain Eric Johnston was dynamic and set the bar high for everyone. All the WRANS and RAN officers, non-commissioned officers, Wrans and sailors I worked with were marvellous. There was a lot of heavy drinking by naval personnel, but the work ethic was outstanding.

Fortunately, the departure of the Task Group didn't mean the end of the RAN's support to the rehabilitation of Darwin. A massive amount of clearing up, demolition, removal of debris, repairs and rebuilding lay ahead. Conditions were stressful, especially when several more cyclones threatened the Northern Territory, and there were strong winds and storms in the first three months after Cyclone Tracy.

The work of the naval personnel from HMAS *Coonawarra* and the patrol boats continued unabated. When I returned to Darwin in late January after my survivors' leave, so much work had been done in my absence and so much had changed, that I felt out of step, out of place. I was issued with new working uniforms and assigned to a squad cleaning out refrigerators and freezers. This was nearly six weeks after Tracy. The houses had been stripped of everything except the fridges and freezers, which hadn't been dumped for some unknown reason, and had been without electricity all that time in summer humidity.

The contents of the fridges and freezers were appalling to smell and feel. We were issued surgical masks, but they didn't help. I spent most of the first three days vomiting and dry retching. With no running water for flushing, there was no point throwing up into the toilets. I often hung over the balconies or out the windows to vomit and breathe clean air, but most of the time I was on my knees, heaving into the buckets we used to clean out the freezers.

To make matters worse, I was the only one sick at the time. The leading Wran of the squad assured me that she and the rest of the crew had been cleaning fridges and freezers for some time and had adjusted to the worst of the ill effects. They were understanding and non-critical. No doubt they had all been just as badly affected when they started. In the meantime, I felt wretched about not being able to pull my weight while they all worked so hard and so well. In later years, when working as an ambulance officer in Rockhampton, I never even gagged over patients' vomitus, faeces or amputated body parts. Nothing was ever as bad as those freezers.

When Wran Radio Operator Special Lyn Alderton returned to Darwin in the third week of February after her survivors' leave, she was assigned to a clean-up crew at Nightcliff. There were two or three Wrans and a few sailors in every team. Fortunately, by this time, lessons had been learned. They headed off in a truck with a list of addresses. Alderton 'expected to see more houses with few walls, lots of belongings strewn around and not much else'. At one address, their assigned house seemed remarkably intact even though the houses on either side were completely gutted. They were lucky to have some very experienced crew members with them when they went into the kitchen. The newbies were told that under no circumstances were they to open the fridge or the oven. Instead, they tied up the handles of the appliances which were then carried out to the dump truck.

* * *

February was a difficult month for me. The fleet had left and those of us who were stationed in Darwin were left to carry on the work. The weather was unrelentingly hot and humid. Tropical rain fell every day with occasional storms. We worked on the clean-up in a ruined city in putrid conditions with no end in sight. Even now, I can cope with heat and humidity for a month or two in summer, but by February my fortitude is at its limit and my family can count on me to grumble about the weather all month.

I finished clean-up duties and was reassigned to communications duties out at Shoal Bay sometime in early March. Returning to comms was an indescribable relief. I loved being a radio operator. Going back to the sets was a deep pleasure, a duty I found even more precious after the events of the past two and a half months. Shoal Bay was even air-conditioned – sheer luxury!

We were only using a small part of the radio station, which was still being fitted out, and most of it was cavernous. A group of us were working away on the radio sets one day, when a sailor hurtled out of another room. He was still sitting on his chair with his feet in the air, the chair's wheels spinning furiously as he propelled himself along by pushing hard off the walls and desks. He wheeled past us at speed and rocketed out the far door. I don't know when and where he finally came to a stop, but a tentative investigation of the adjoining room revealed a snake which had slithered up one of the cable outlets beside him. Something of a panic ensued.

In our infrequent downtimes, we would play cricket in one of the unoccupied sections, using scrunched-up paper wrapped in masking tape for a ball and planks of wood or rulers for bats. In later years, that area was filled with rows of radio sets, and hummed with activity.

On 4 April 1975, Major William Hughes reported for duty in Darwin. Commanding officer of HMAS *Coonawarra*, Stan Hall appointed him officer in charge (OIC) Shoal Bay Communications Station. The fitting-out of the station continued around us while we worked, and we continued to adapt to changing conditions.

* * *

The local Navy vessels remained busy. Repairs and maintenance were ongoing, and normal patrols had to be undertaken. The patrol boat HMAS *Advance* had been slipped on Friday 17 January for inspection and cyclone-damage repairs. All possible repairs were undertaken, but it was decided that the bent 'P' bracket was a major defect that required attention in Cairns, and *Advance* was unslipped on Sunday 26 January. On Friday 7 February, HMAS *Advance*, the patrol boat HMAS *Barbette* and the LCH HMAS *Tarakan* departed Darwin for Cairns. *Advance* underwent repairs in Cairns for three weeks, leaving Friday 28 March for the journey back to Darwin.

The patrol boat *Assail* was slipped on Friday 31 January for two weeks' survey and repairs. On Tuesday 18 February, HMAS *Assail* departed Darwin Harbour on a patrol, which was cut short a few days later by bad weather. *Assail* sailed again on patrol on Monday 24 February. Three weeks later, she took three Indonesian fishing vessels into custody. *Assail* and the patrol boat HMAS *Aware* commenced towing the vessels to Koolan Island, but one of the vessels, which had been extensively storm damaged, sank en route. The Indonesians were prosecuted and found guilty of fishing in Australian waters, but no convictions were recorded. *Assail* and *Aware* were tasked with towing the Indonesian vessels back to sea and releasing them. The justification for that decision was inexplicable, and the experience was demoralising for the patrol-boat crews.

The patrol boat HMAS *Adroit* arrived in Darwin on Sunday 16 March as HMAS *Arrow*'s replacement, but battery problems prevented her from undertaking immediate patrols.

In May and June 1975, the minehunters HMAS *Curlew*, HMAS *Ibis* and HMAS *Snipe* surveyed the approaches to Darwin and the harbour itself, locating trawlers sunk during Cyclone Tracy and other navigational hazards.

* * *

The living and working conditions of naval personnel and their dependants in Darwin continued to be challenging. For months, civilian technician Michael Pilkington divided his time between Shoal Bay and his house in Casuarina.

He remembers his neighbour Jock McCallum saying, 'Listen, mate, I've got to get my wife back.'

Pilkington nodded and said, 'Fair enough.'

Pointing to a concrete slab in the Pilkingtons' yard, McCallum asked, 'If I can get a caravan, can I have that?'

Pilkington readily agreed. Jock got his caravan and set it up for him and his wife. Pilkington's next project was reroofing his house. Eventually, power poles were erected, and progress was being made to restore electricity along their street.

'As soon as you had power – Woohoo!' says Pilkington, grinning broadly. The first thing he and McCallum did was dry out all the ceiling fans. They got them working and reinstalled them. Once the house was reroofed, he sent a message to his wife, Christine – 'Palace reroofed' – which meant she and their boys could finally come back to Darwin.

They got some bunk beds for their boys but, apart from that, Christine says the family's living standards were 'pretty rough'.

And they had no bedrooms. 'We were just doing indoor camping,' she says. 'We had plastic ceilings. You could watch the mice running around up there while you were in bed.'

Michael and Christine had applied to buy the house they were living in, but the application hadn't been processed. There were rumours that the government was going to take the property from them, and they didn't know what was going to happen. They waited for about sixteen months.

'So, we just sat there in this wreck of a place,' says Christine. They were at least given a caravan and were able to put down some floorboards for an annexe, where they 'were quite comfortable'.

Eventually, inspectors came and checked the stumps and beams of the house and found no real damage, and the Pilkingtons were told they could rebuild the house with the existing stumps. 'So that's what we decided to do,' says Christine. They continued to live in the caravan while rebuilding the house.

Their friend Jock McCallum was well known for playing the bagpipes, and he would stride up and down, playing them, while the Pilkingtons' house was being rebuilt. He claimed he had to keep his practice up. Christine says that everybody who was still in the street loved it.

* * *

Living conditions at HMAS *Coonawarra* were still very poor by the end of March 1975. In his report of proceedings, the commanding officer, Stan Hall, reported that sailors were sleeping on 'the ground floor of their barracks, which lost the top half of the building, to ad hoc dormitory cubicles in the Amenities Block'.

'Most,' Hall reported, '[were] sleeping in double-tiered bunks obtained by welding together beds salvaged from married quarters.'

Several houses in the married quarters had already been demolished since Tracy. Many other buildings were reroofed by the shipwright team, which had been posted in for the rehabilitation work. A reroofed cottage was being used as the wardroom. Wrans officers were accommodated on the top floor of the senior Wrans block. Commander Hall further reported, 'Satisfactory progress is being maintained re rehabilitation and reconstruction. Forty-two houses are more or less weatherproof. Thirty-eight are allocated as married quarters and four as single quarters. Most houses are dirty with mud and mildew.'

Despite the injuries he sustained during Cyclone Tracy, Petty Officer Mick Hamilton was sent back to Darwin.

He says, 'I was discharged from *Penguin* around Easter time, I think, and they declared me fit, so what did they do – send me back to Darwin!'

His previous job at HMAS *Coonawarra* no longer existed, so he was made the depot electrician and tasked with connecting power in the marriage patch at *Coonawarra*. Many houses were reduced to floorboards on stilts, and caravans were parked on the concrete slabs beneath. Hamilton wired the vans up for electricity. He 'felt it was extremely unfair for a sailor in his little fifteen-foot van on one side and an officer in his twenty-five-footer on the other side to share the cost of electricity'. He also didn't think anybody should be paying for electricity at the time. He wired up the vans and gave them power but didn't connect them to the meters. He thinks he wired up about thirty before it was noticed that the occupants were receiving free power. Hamilton was not put on report. He suspects the authorities 'understood that it was ethically wrong to send back pre-cyclone people to do this sort of work'.

Not surprisingly, Hamilton's physical condition deteriorated. He was sent south to the communication base HMAS *Harman* in Canberra, and his injuries were treated by an orthopaedic surgeon.

* * *

Bernie Kloss was also sent back to Darwin. He was in Cairns with *Attack* until Easter, when *Advance* arrived in Cairns to get its propeller fixed. The second engineer on *Advance* was married, and as there was no accommodation for married people in Darwin, Kloss was told by his skipper to 'pack [his] seabag again'. He was 'to swap out with the second engineer on *Advance* and go back to Darwin'. Kloss had only been in Cairns for a month and wasn't too thrilled about the prospect of returning to Darwin.

Back in Darwin, Kloss reckons they were at sea eighty per cent of the time, with only two patrol boats now on the lookout for foreign boats illegally fishing in Australian waters. '*Arrow* was gone, and it would be fourteen months before *Attack* was back in action again,' he says, which meant *Advance* and *Assail* 'were at sea fulltime'.

As a single man and second engineer, Kloss's tenure in Darwin was supposed to finish partway through 1975, after a twelve-month posting. When the time came, however, Kloss was told the Navy couldn't find any volunteers to go to Darwin, so he was asked to extend his posting by another twelve months.

Sometime during Kloss's extended posting, *Advance* sailed out past Broome to Rowley Shoals on another fishery patrol. They dropped anchor for the night and decided to have a movie night. Sailors rigged up a couple of broom handles on the back deck and spread a white sheet between them for a screen.

They set up the 16mm projector and, just before they started the movie, their CO said, 'I've got a little duty that I have to carry out here first.'

They all glanced around and said, 'Yeah, what's up, boss?'

'Petty Officer Kloss, now I believe?' he said.

'Yes, sir.'

'I have something here from Naval Office for you.'

'What's that, sir?'

Kloss was awarded a Naval Board Commendation, along with Able Seaman Danny Dade, who had connected HMAS *Attack* to the cyclone mooring buoy in Darwin Harbour on Christmas Eve. To this day, however, Kloss believes the reason the *Attack* crew were 'able to sit on the rocks, have a few beers and throw chicken bones over [their] shoulders on Christmas Day' was Leading Hand Stoker Kenneth John O'Hare, whose nickname was 'Rabbit'.

'He was the one, rough as it was,' says Kloss, 'who kept going down in the engine room during the night, checking the engines, making sure they were okay, making sure the air bled off the cooling systems and that kind of stuff.' Throughout the night, O'Hare would step out of the engine room and shout up the short stairway to the wheelhouse, 'Hey, Bernie.'

'Yeah.'

'Everything's alright down there, mate.'

'Thanks, Rabbit.'

'And that was it,' says Kloss. 'But he was the most important guy maintaining the machinery all night. He did a good job.' Kloss only saw him once after that, a couple of years later in Cairns. He doesn't know what O'Hare did after leaving the Navy.

* * *

Conditions at HMAS *Coonawarra* were still sub-standard by the end of June 1975. The debris had been removed from the establishment, although there was still some littering the surrounding bushland. Buildings beyond salvaging were being demolished and the rubble removed.

The clear weather of the dry season meant work could go ahead without delays due to rain, and the third chippies team on rotation in Darwin performed consistently high-quality work. There were delays, however, as some sections of the civilian workforce in Darwin went on strike over their conditions.

In mid-1975, Commander Stan Hall was still concerned about the crowded accommodation for senior and junior sailors at *Coonawarra*. Some old married quarters (circa 1938) had been allocated to 'unaccompanied sailors', but the commander had trouble understanding 'why the 100-man accommodation block originally planned for the 1975/76 works program' had been deferred. Hall believed the morale of junior sailors accommodated at *Coonawarra* would suffer if they had to live in such crowded and sub-standard conditions indefinitely. Fortunately for us, Wrans accommodation was good.

Of the married quarters, which were 'critically short', the numbers speak for themselves. Of the ninety-eight married quarters within *Coonawarra* West, *Coonawarra* East and Hidden Valley Estate: twenty-eight were completely destroyed; forty-one rehabilitated and reoccupied by families; eight were used for single unaccompanied sailors' accommodation and two for unaccompanied male officers; one as wardroom; and eighteen remained to be rehabilitated or reconstructed, with an estimated further year's work to make them liveable again.

Consideration had to be given to upgrading building standards on the base to withstand future cyclones. Even two of the solid-construction buildings, the Wrans quarters and senior sailors block, were acknowledged to be hazardous during high winds. During the cyclone, the walls of the new Wrans quarters held up well, but the breezeway design made the movement of personnel through the building dangerous, something I experienced firsthand. The senior sailors block was deemed hazardous because of the large areas of glass used in its design.

We were dreading the next wet season. In the meantime, we made the best of the good weather. The second annual

Beer Can Regatta was held on Sunday 15 June and was again a huge success. The boats had to be made almost entirely out of empty beer cans and the ingenuity and craftsmanship of the entries was outstanding. Some entrants didn't take the sailing of the craft very seriously and entered for the fun of tossing flour and water bombs at other entrants. It was a wonder some of the craft floated at all.

Our wonderful fellow Wrans made us proud with the Navy Darwin entry. It was a twenty-foot, three-masted barque, designed by Chief Wran Nerida Sullivan and built by a team primarily made up of Wrans. It was rumoured between 4500 and 5000 empty beer cans were used in its construction, but who's counting. The boat won first prize for 'the most ingenious' craft. Then the crew of Wrans sailed her to third place in the sailing section.

The glittering Darwin Navy Ball was held at the Don Hotel on 12 September.

Exercise WARM-UP was held in October, testing the preparedness of Darwin for another cyclone. The local Passive Defence Organisation of HMAS *Coonawarra* was also tested, and a lecture about cyclone preparedness was given to married personnel and their wives. Many naval families were still accommodated in caravans, so a demonstration was given on securing a caravan in high winds. After that, forty-two wives and children were transported to the Shoal Bay Communication Station, a designated refuge centre for *Coonawarra*, testing the family reception and welfare organisation.

It took to the end of 1975, a year after Cyclone Tracy, for the structural rebuilding of the junior sailors block to be completed and the building waterproofed. Even then, it wasn't fully occupied until the electrical rewiring was completed in early 1976. Naval and victualling stores were finally located in leased weatherproof buildings.

Most of the Navy buildings that were unsalvageable due to extensive damage from Cyclone Tracy were demolished and the debris removed by the end of 1975. These included many of the houses on the base, as well as the old transmitter building and masts. However, at the Shoal Bay Communications Station, the water tower, diesel house and water tanks, which were damaged during Cyclone Tracy, still hadn't been repaired.

As well as the sub-standard accommodation issues for officers, junior rates and families, there were pay problems as well. The new rates of district allowance for naval personnel in Darwin had been set on 2 July 1974, and we were also eligible for a Darwin disaster allowance. Commander Hall reported that, by 31 December 1975, the statutory approval for payment of these allowances was still being delayed.

* * *

Bernie Kloss had been serving in Darwin since July 1974, firstly on HMAS *Attack* and then on HMAS *Advance*. After his second year in Darwin, the *Advance* crew were asked to stay on there. In 1976 they still hadn't had any annual leave but were asked to stay on for yet another twelve months. Kloss agreed. He was enjoying his service, working hard and having fun.

When *Attack* returned to Darwin toward the end of 1976 'repaired and all good again', the postings were changed, and Kloss was sent back to *Attack* as second engineer. He was there for another four or five months, and 'went up and around New Guinea, did a couple of patrols up that way'.

Attack was alongside in Darwin doing some engine servicing when the coxswain came back from a visit to *Coonawarra* and asked Kloss how long he'd been in Darwin. Kloss replied he'd been there since July 1974. The coxswain asked if he had any leave; Kloss said 'no' and explained why. The coxswain told him he was owed six months and one week leave on the spot. After four years with no leave, Kloss said 'that sounded about right'.

* * *

Navy cook Michael Daly also continued his association with Darwin. He stayed on HMAS *Advance* until early 1976. He then had to have a shore posting, so he requested to go to HMAS *Coonawarra*. He was happy to be posted there and spent another twelve months at *Coonawarra*, which was 'like home', he says.

Daly was accommodated in the junior sailors block for a week or so before moving into one of the vacant married quarters on the base. The house was on the corner opposite the gangway, where personnel caught the bus out to Shoal Bay.

He says, 'I lived in that house with a good mate and cook, "Ferret" Brennan. There were a couple of other blokes in there. We had an absolute ball.'

Daly also loved working out at Shoal Bay. 'It was great.' Guard duty was still hazardous because of the wildlife though. Daly says the sentry on the gate 'had big gumboots and they made him a wire rope attached to a chunk of wood, so every time he did his perimeter patrol, he could belt the snakes'. Daly himself 'would never walk outside at night-time'.

The cooks at Shoal Bay would order 'a flagon of plonk' to drink every night and the victuallers would ask why they needed 'so much bloody wine in the food'. Drinking was used as a coping mechanism. When not on duty in the evenings, most of us had a drink or three at a smokie or at the Rates. Sport was important, and the cinema and drive-in were popular, but drinking was prodigious.

Daly says, 'People were really messed up at *Coonawarra*.' He recalls 'vividly' the shooting death of a young Wran – 'a really nice girl' – by a sailor.

'We were up the pool,' he says, 'having a really big smokie', when the sailor 'came up to her at the pool, shot her, and then killed himself, outside the entrance to the pool'. He describes it as a 'horrible incident' that 'added to many people's issues'.

* * *

I stayed in Darwin until November 1975, when I went on leave and was then posted to HMAS *Lonsdale* in Melbourne. We Wrans were accommodated at Greenwich House in Toorak. That was a fabulous experience, and I loved living and working in Melbourne. After only a year, though, I was posted back to Darwin, starting work at Shoal Bay in December 1976, just in time for the cyclone season again. This posting lasted until June 1978, when I received another posting to Melbourne. I paid off from the WRANS when my term of enlistment was up in February 1979.

CHAPTER 29

CARRY ON

In *The Lord of the Rings* by JRR Tolkien, the hobbit Bilbo considers finishing his book with the words, 'And he lived happily ever afterward, to the end of his days.' This book can't end like that.

Many, if not most, of the naval personnel who endured Cyclone Tracy and the clean-up remain affected by their experiences. Ruth Turnbull stated in 2015 that for years, she had trouble sleeping. 'I used to wake up,' she says. 'I couldn't go back to sleep. Things would just go around and around in my mind. And I still dream about Tracy to this day. My psychologist helped me over a hell of a lot.'

Ruth paid off up in Darwin and said, 'Tracy was the worst, but then after Tracy we had Max, Gretel, Ingrid, Thelma and a couple of others, and going through all that would just take me straight back. I was a nervous wreck in most things.'

Ruth became a radio operator for the civilian Radio Communications Group, working out of the Victualling Yards in Geranium Street, Stuart Park, Darwin. Whenever cyclones threatened Darwin, her colleagues made sure she stayed safe with them in that solid brick building.

Our dear friend Joan Czarnecki says, 'Things sort of returned to the new normal before I headed off to HMAS *Albatross* at Nowra in April 1975. Those are the things that I can still recall, the timeline may not be totally accurate, but to this day I still remember.'

Mick Hamilton still struggles with the injuries he sustained at the old transmitter station when the building collapsed around him. He suffers from sciatica and has permanent curvature of his upper back. He has undergone multiple laminectomies, has steel rods in his back and sees a physio twice a week. Hamilton says he's 'getting over the experience of Cyclone Tracy quite well now' because he's 'accepted the situation', even if he does not 'accept some of the decisions that were made by officers at *Coonawarra*'. He credits his and his wife Doreen's survival to being at the transmitter station and not in their flat in Casuarina, which was demolished by Tracy. He still gets flashbacks to the cyclone and manages his PTSD with medication.

'Sometimes,' he says, '[I] get depressed for no reason at all, and I have dreams every night – they're all Navy dreams – every night. I don't have dreams of my life after the Navy at all. Never. And I had very stressful jobs afterward, but never dreams about them.'

Liz Grimes, who served in the Navy for another eight years after Tracy, thinks many 'still suffer from the effects of the wind and noise'. She thanks her fellow petty officers at HMAS *Cerberus* for their patience in helping her through the aftermath. She says the impact on people like her 'who went through the cyclone – the effects of gale-force winds – were felt for months, sometimes longer'. She remembers returning to Darwin on leave for the first time since Tracy and the locals praising the efforts of the Navy in the clean-up.

'I was ever so proud to be in the Navy at that time,' she says.

For Carolyn Spencer, Tracy is 'never far away' in her mind.

'Wind or lightning, or just the name, brings a lot of emotion to the surface,' she says, 'but despite all that we still have our lives.'

Michael Daly of HMAS *Advance* recalls going to see the doctor when *Stalwart* arrived in Darwin. He says, 'Our debrief was, "Are you okay? Have you got any injuries?"

'"No."

'"Alright. Great. See you later."

'That was it!'

'Of course, a lot of what happened didn't hit home then, but it did later,' Daly adds. He would 'get a bit concerned in particularly fearsome weather at sea sometimes. You'd get stuck in a cyclone over in the West, Queensland or Southeast Asia and you'd have to go through one of them again, but I never went through anything as strong as Cyclone Tracy.'

Daly stayed on *Advance* through 1975, stopping in Cairns for a refit of the patrol boat in March, where he celebrated his twenty-first birthday. He bought a bottle of tequila and a pound of lemons, went to the back of his accommodation, and spent the evening drinking with a mate. Working and drinking was 'pretty much all we did for the next twelve months,' says Daly. He admits, 'we self-medicated and never got any sort of debrief or counselling. I've never seen so many people drink so much grog in my entire life as they did in Darwin.' Daly spent thirty-three years in the Navy, but 'as much as [he] loved it', the Navy ruined him 'health wise'. These days, he feels he's 'dealt with pretty much everything':

> It's something you went through, and you strived to do your best and you kiss your arse every day and thank God you made it out. I've got so many Navy mates that died in the last two years – all in their sixties. Mainly cancer. Bloody cancer. Tracy and her aftermath left a lot of damaged people.

Like other Navy personnel, Bob Dagworthy and the other crew members of *Arrow* did not receive any counselling or psychological support after Tracy either, despite the loss of their ship and two shipmates. As an officer, Dagworthy believes he 'was entrusted to be in command of a ship and expected to get on and deal with any circumstances [he] experienced'. He doesn't think that expectation helped him cope, either at the time or later on, and believes that's the case for many others who were in Darwin during Cyclone Tracy.

After his posting to *Bombard*, Dagworthy got a shore posting, then headed to Fiji to command the Fijian Navy's senior ship, HMFS *Kikau*, when their Navy was starting out. He was forced to avoid three cyclones while serving in Fiji. As the senior officer for the Fijian Navy's three ships, Dagworthy's policy was to put the ships to sea when cyclones threatened. 'I treated every cyclone as another Tracy,' he says.

Later, Dagworthy was appointed commander of the Patrol Boat Force, CO *Waterhen*, and then commander of the Australian Mine Warfare and Patrol Boat Forces. When personnel were posted to serve on patrol boats or in command, he says, 'whether it was in Darwin, Cairns or Sydney', he would stress to them that 'if a cyclone is approaching, they must not remain in harbour'. On retirement, Dagworthy had a hip replacement from the injuries he'd sustained when he 'washed through the wharf in Darwin'.

After leaving Darwin, Dagworthy's wife, Christine, didn't receive any counselling either. There was nothing to help her support her daughters, Siobhan and Kate, or her husband 'with the trauma he was experiencing'. She describes their family as very resilient people who learned to cope when plans went awry. As a Navy wife, Christine had her share of challenging postings and long separations. Siobhan was six and Kate was four when the family was transferred to Fiji. When cyclones threatened the country, 'Bob was away at sea in his role with the Royal Fiji Navy. The memories all came back to me,' says Christine. When trees started falling down around the house and along the driveway, Christine thought, *Oh goodness, I'm going through all this again.*

The Dagworthys went back to Darwin briefly in 1994 and 2007 and visited the graves of Les Catton and Ian Rennie. Bob and Christine returned again in 2017 in a motorhome after travelling through Queensland and the Northern Territory. Arriving in Darwin, they tried to locate a stained-glass window that had been commissioned to remember the *Arrow* crew. They searched every church in Darwin and the complex that had been HMAS *Coonawarra* without success. They were later advised that the window was stored on Spectacle Island in Sydney Harbour for safekeeping.

The Dagworthys have since been in touch with the Northern Territory Museum in Darwin and gifted them a ceremonial lifebuoy from *Arrow*. A Navy clearance diver had recovered it and given it to Bob. The museum now has a display featuring 'a photo of *Arrow*, the cyclone buoy and a mention of the two sailors who lost their lives', says Christine. 'There have been times when you think you should write these memories down,' she adds, 'but it's painful. Emotionally painful. It has been difficult to revisit what happened to *Arrow*. You live with survivor's guilt.'

Bob Dagworthy is not critical of the Navy's actions and policies of the day. He says, 'the strength of Cyclone Tracy was unknown' and they believed 'the cyclone buoy arrangement was appropriate at the time'. They've 'since learned many lessons, and improved procedures for managing vessels in cyclonic conditions'. As the former commander of the patrol boat forces, Dagworthy has 'witnessed vast improvements with every new class of patrol boat – from the Attack class to the Fremantle class and now to the Armidale class'. The current patrol boats, he says, are 'also larger and far more comfortable'. As well as stabilisers, they're 'equipped with modern, appropriate armament. Not a Bofors 40/60 that previously served in an anti-aircraft role in London during the Second World War.'

* * *

My behaviour, like that of many of my fellow servicemen and servicewomen, altered significantly after the event. Many took to alcohol and risky behaviour. I'd only had an occasional drink prior to Tracy, but afterward became a heavy drinker for years and displayed poor judgement in personal matters.

For decades afterward, I suffered from nightmares. Going to sleep was easy but staying asleep was difficult and I would often wake with a yell of fright.

The careers of many of us stalled. I had planned to apply to become an officer as soon as I turned twenty-one, but never progressed beyond leading Wran (equivalent to an Army corporal). Over the next four years, numerous Wrans who had less service than me jumped past me in rank. I didn't realise that I had Post Traumatic Stress Disorder (PTSD) but knew I didn't have the confidence anymore to strive for promotion.

Other Wrans who went through the cyclone simply left the service quietly, long before their period of enlistment was up. This was quite common, in fact, but I strongly suspect the rate of resignation of veterans of Cyclone Tracy and the clean-up was higher than normal.

I went on to serve four more years in the WRANS and another five years in the WRANS Reserves. I had received the award for being 'The Most Outstanding WRANS Recruit' in my intake, but after Cyclone Tracy my career was less than stellar. I had a drinking problem for a few years, lost my drive, and was aimless until the end of 1981, when I took charge of my life again. I then pushed myself too hard – in part because of the sentry incident.

In his report of proceedings, January–March 1975, NOCNA Captain Eric Johnston noted, 'I took advantage of the presence of the fleet medical officer and his staff to have all personnel examined for possible physical or psychological effects from the cyclone.'

I don't recall undergoing any such examinations and there were certainly no follow-ups. In 1978, while serving again at HMAS *Lonsdale* in Melbourne, the unsympathetic civilian doctor employed by the RAN referred me to a civilian psychiatrist. I was struggling and didn't know what was wrong with me. I never talked about Cyclone Tracy in those days, and the subject was never raised by those medicos, no doubt because it wasn't mentioned on my service records, so I, along with so many others, was left to deal with undiagnosed PTSD by myself.

Looking back, it seems to me that nothing much had been learned since the destroyer HMAS *Voyager* sank after colliding with the aircraft carrier HMAS *Melbourne* in 1964. *Voyager* survivors, who had lost shipmates, their ship and their possessions, were granted a week's leave before being posted to another ship. They received minimal post-collision mental care or recognition of trauma.

There was a nameless constant pain deep in my chest that wouldn't go away. However, I had a son, Terence, of whom I am immensely proud, and put him through a private school and tertiary studies. He has always been my anchor and my compass. Being his mother has been a joy.

A few years after leaving the WRANS, I became an assistant technician with Telecom Australia and worked as a telephone installer. I was the first female telephone installer in my town, but the technicians and linesmen were great fun to work with and the bosses were encouraging. They ensured that I was treated the same as the men. When I joined the ambulance service, however, I found that the attitude toward women in male-dominated professions was vastly different there.

As an ambulance officer in the 1980s and 90s, I was subjected to daily put-downs from many male officers at my station who disapproved of having women in the service. It seemed to me that management, up to and including the assistant commissioner, failed to address issues of sexism and bullying, and even contributed to the problem. Personal insults and bullying are so common in many workplaces that they're considered normal, so there was nothing special about my situation, but I handled it badly. I wouldn't tolerate being put down by anyone and fired up whenever I was given a hard time. I freely admit that this made matters worse for me.

An example of the discrimination was the sleeping conditions at the ambulance station. During the night, when not required for callouts, the four ambulance officers on duty were permitted to sleep. The men had the choice of two dedicated bedrooms, both with two beds, good mattresses, pillows and linen. They were internal rooms with no windows and could be completely darkened.

By contrast, we three female ambulance officers had a fold-up camp stretcher in the Air Ambulance pilots' office, with no curtains on the window and a security light on all night outside. When the pilots moved out to the new hangar at the airport, the room was left vacant, except for the camp stretcher.

When I qualified as an ambulance officer after three years of on-road work and training, I told my colleagues that I was no longer going to put up with sleeping on the camp stretcher. I slept in one or other of the dedicated bedrooms, and if my working partner at the time didn't like sharing the room with me, he was welcome to sleep on the camp stretcher in the women's room.

That didn't go down well. The men were not comfortable sharing a bedroom with me and refused to sleep on the camp stretcher. The excuse of 'budgetary constraints' had denied the requests of female ambulance officers for a dedicated bedroom for many years. However, when the male ambulance officers agitated for the women's room to be upgraded, management promptly complied. Money was found in the budget and, within three weeks, the former pilots' office was fitted out with a proper bed, curtains were hung on the window, and the men had their bedrooms to themselves again. There was, of course, a backlash against me, but at least we women were finally able to sleep in comfort.

Another problem existed with the showering arrangements. After particularly nasty jobs, we needed to shower and change into clean uniforms. We female ambulance officers had to use the men's shower, which was situated in their toilets. I was told, 'If you can't take the heat, get out of the kitchen,' and 'That's just the way things are, the way they've always been. If you don't like it, you can leave.' I lost count of the number of times male ambulance officers said to me, 'You have to earn my respect.'

Some of the male ambulance officers were wonderful, and I maintain the greatest respect and liking for them. There weren't enough of them, however, to mitigate the negativity of the others. I absolutely loved patient care and giving people valuable medical care when they were at their most vulnerable. However, I hated the culture of the ambulance service and was a disaffected employee, feeling constantly and unrelentingly angry.

I found myself fighting a battle on two fronts: the present one against discrimination and harassment, and the second against the memories of Cyclone Tracy and the clean-up. I felt that I had done enough during and after Cyclone Tracy to warrant respect. The fact that the Navy didn't consider our work in Darwin during Cyclone Tracy and the clean-up even worth noting on our service records, however, had me doubting that and hurt deeply. Maybe my little contribution wasn't significant. Maybe I wasn't significant. I should have done more, worked harder. I should have behaved better. The memories crowded in and the nightmares, the cold sweats and the anger increased.

After sustaining a back injury, I was particularly vulnerable and eventually had a breakdown. Only two things prevented me from attempting suicide. The first was my son, who I love with all my heart. He is a remarkable fellow, and I wanted to be around to support and encourage him as he grew into adulthood, as he deserved.

The second is the fact I am a stubborn bitch and refused to give up. The breakdown resulted in my being diagnosed with PTSD. Finally, the leaden pain that had lived deep in my chest for over twenty years was given a name and I knew what I was dealing with at last.

Transferring to admin work with the Queensland Police Service was the saving grace for me, as the police officers and their management were wonderful to work for. They gave me back some faith in people. Unfortunately, I was still a difficult employee. I worked hard but wasn't easy to work with, as managing my PTSD grew increasingly difficult. Another breakdown in my fifties resulted in me having to give up full-time work.

* * *

I underwent various tests and interviews leading up to the diagnosis of PTSD. A few perplexing questions were asked about my memory. One was, 'Have you forgotten anything about the event?'

Answer: Yes.

The next was, 'How much have you forgotten?'

Really? How on earth would I know what I've forgotten if I've forgotten it? I only know that some incidents are gone from my memory when one of my friends mentions an event and says that I was definitely there at the time, standing beside her.

Goodness knows, I would feel a lot more comfortable if I had no memory of some traumatic or embarrassing events, but life doesn't work like that. Some of the forgotten memories are of fun times. Take water polo, for example. When my son Terry was in primary school, my mother and I religiously followed the Olympic and Commonwealth Games, and we would always have a globe of the world and a large atlas on hand so that we could look up where the competitors came from. After all, where on earth is Tobago?

Mum and I also used to do Roy Slaven and HG Nelson routines even before we knew about Roy and HG. We'd watch a competitor in the high diving do a double somersault with pike, and Mum would scoff that she was 'doing triple somersaults with reverse pikes in primary school!' (She couldn't even swim.) While watching the basketball competition, I would claim to have slam dunked basketballs while in Grade Three. (I played netball at school, never basketball, and I wasn't tall.) Terry would inevitably close his eyes, bow his head and thump his palm on his forehead in resignation at our outrageous claims, and we'd crack up laughing.

One day when I came home from work, Terry and Mum both rushed me with tales about the water polo matches. They were agog at how violent the male and female competitors were. When we watched a match peppered with savage wedgies, punches, kicks and attempted drownings, I scoffed and said that when I played water polo in the Navy, we were much worse. There were bashings and blood in the water from start to finish, with ambulances lined up at the exit. (As if!)

When my mother passed away years later, Terry and I were going through paperwork, letters, cards and other memorabilia prior to selling the house. I came across some small diaries I had kept while in the WRANS. During my second posting to Darwin in 1977, I was alarmed to discover one day that I had been invited to three parties in one night. I attended all of them, but it was quite a stretch, so I decided to keep diaries to make note of my watches, duties and social events to avoid that ever happening again.

While thumbing through these small diaries with Terry, I noticed an entry for 'water polo training'. Surely not. Then there were a lot more entries for water polo training and matches. No memory of that at all.

I rang up my friend Ruth Turnbull and asked, 'Ruthie, did you know that I played water polo in the WRANS?'

She replied, 'Of course. I used to go to all your matches and cheer from the sidelines.'

Apparently, gaps in memory like this are part of PTSD. You carry on with life as usual, working, socialising, playing sport and so on, but for some reason there are blanks – some memories just don't stick. That's disappointing because I would have enjoyed playing water polo.

If I concentrate hard, I have a flash of a memory of being in the water and holding a ball in the air ready to throw it, but that's all. It's unlikely that there would have been wedgies, bashings and blood in the water from start to finish, with ambulances lined up at the exit, but it would have been fun anyway. Imagine the delight in our household at discovering Roy and HG when my son was in high school. Their later commentary at the Sydney Olympics was brilliant, and we were again agog at the violence in the men's and women's water polo matches.

Today, I owe a great debt to the Vietnam Veterans Association. The work they did to ensure their fellow Vietnam veterans received recognition for their service and support for their mental and physical health was outstanding. I don't have the words to express how grateful I am that they then made that support available to all veterans. They, the Department of Veterans' Affairs, my wonderful family and loyal friends deserve the credit for the quality of life that I experience now. A lot of the hurt has thankfully abated, but the emotional scars remain.

Even so, I never felt that I was a victim of Cyclone Tracy. We Wrans survived an extremely dangerous cyclone, lost all our possessions, suffered physical and/or psychological injuries but, to me, we were never helpless victims. We got down to work straight away. We were rescuers, clean-up crews, salvage teams, demolition crews, cleaners, pest controllers, sentries and couriers. We helped set up communications at Shoal Bay. We assisted civilian relief organisations and manned communication centres for the police, the fire brigade and the Department of the Northern Territory. We did good work and I have always been proud to belong to that little band of Wrans. That experience continues to spur me to live an intentional life.

A few years ago, Ruth Turnbull and I recorded two of the conversations that we enjoyed together when I spent a last weekend with her. We were practically incoherent with laughter the whole time and most of the stories about our life in the WRANS were only half told before they reminded us of another incident, and we were off again!

Ruthie treasured every minute she spent in the WRANS, and her memory was remarkable. The funny situations, the sad, the tragic – all her memories were important and precious. Just because she's passed doesn't mean her treasured memories should fade away. And that applies to all of us.

During a report on ABC TV in 2017, Bob Dagworthy mentioned that he was worried people are beginning to forget what happened. 'It's important to talk about it. To state the facts as they really were,' he says. 'I want people to remember what happened to *Arrow*, and I want people to remember the two sailors who lost their lives.'

We should never forget the sailors who died, and the heroism displayed by so many other sailors and Wrans. We won't forget the mateship, the stoicism and the determination to survive, to work, to clean up and to rebuild. That's what continues to give us strength.

Lest we forget.

NAVY 'SLANGUAGE' AND TERMINOLOGY

Buffer: Sailor in charge of the boatswain branch. As the executive officer's right-hand man and the one by whom he passes orders, he is considered to be the buffer between officers and ratings

Bulkhead: A wall

Cabin: Accommodation room

Chippy: Shipwright (carpenter)

Deck: The floor or ground

Deck head: Ceiling or roof

Dhobey: Having a shower or a wash

Dog watch: The watches from 1600 (4 pm) to 2000 (8 pm) daily

Fathom: A measure of depth equal to six feet

Flying bridge: An open area on top of a surface ship that provides unobstructed views around the vessel

First officer: The executive officer. Also known as the XO

Forecastle: The deck at the front of a vessel

Galley: Kitchen

Gangway: Any recognised entrance to a ship or establishment, or a passageway within a ship. Also used as an order or warning to get out of the way

Greenie: An electrical sailor

Hove to: In a stationary position with head to the wind; at a standstill

Junior sailor: Those with the rank of ordinary seaman, able seaman and leading seaman

Kellick: A leading seaman, equivalent to an army corporal

Kit: Uniform

Knot: A measure of speed, with one knot being equal to one nautical mile per hour

Limers: A sort of powdered cordial, usually served chilled on ships at sea
Lower deck: All ranks below that of officer
Mess deck: Sleeping and recreation area in a ship
Paid off: Left the Navy at the end of the period of enlistment
Pennant number: A ship's unique identifier, painted in white on the hull
Port: The left-hand side of a vessel (the navigation light is red)
Pusser: Anyone in the Navy or the supply officer (derived from purser)
Pussers: The Navy
Quarterdeck: The stern deck area of a vessel
Rig: Sailors' term for uniform
Rounds: Nightly or other inspections of mess decks or accommodation
Scran: Food or meals
Run in: Being charged by naval authorities for breaking the rules
Scuttle: Porthole or window
Scuttlebutt: Rumours whether true or false
Secure: Cease work for the day
Senior sailor: Those with the rank of petty officer, chief petty officer or warrant officer
Skipper: The captain of a ship
Smokie: Party
Starboard: The right-hand side of a vessel (navigation light is green)
Stoker: A marine technician, one who works in the engine room
Wardroom: Officers mess
Watch: Period of work, e.g., the day watch is from 0800 (8 am) to 1600 (4 pm)

REFERENCES

Cabban, Peter and Salter, David 2005, *Breaking Ranks*, Random House Australia

Cole, Keith 1977, *Winds of Fury– The full true story of the great Darwin disaster*, Rigby, Adelaide.

Department of Defence 1979, *The Defence Force in the Relief of Darwin after Cyclone Tracy*, Historical monograph No. 15, September, Directorate of Instructions, Orders and Manuals, Administrative Services Branch.

Department of Science/Bureau of Meteorology 1975, *Cyclone Tracy: An Interim Report by Staff Members of the Bureau of Meteorology*, Technical report 14, Australian Government Publishing Service, Canberra.

Durisch, Peter, Howarth, Robert (Bob) & Stevens, Kenneth (Rick) 1975, 'Cyclone! Christmas in Darwin 1974', *Sydney Morning Herald*, John Sands, Artarmon, NSW.

HMAS *Advance* 1974, *Reports of Proceedings, HMA Ships and Establishments*, File no. AWM78 41/4, January-December.

HMAS *Assail* (I) 1974, *Reports of Proceedings, HMA Ships and Establishments*, File no. AWM78 41/4, January-December.

HMAS *Attack* 1974, *Reports of Proceedings, HMA Ships and Establishments*, Item no. AWM78 43/5, January-December.

HMAS *Betano* 1974, *Reports of Proceedings, HMA Ships and Establishments*, Item no. AWM78 64/1, February-December.

HMAS *Coonawarra* 1975, *Reports of Proceedings, HMA Ships and Establishments*, File no. AWM 78, January-December.

Johnston, Eric (Commodore) 1986, *Operation Navy Help – Disaster operations by the Royal Australian Navy post-Cyclone Tracy*, address delivered at the State Reference Library of the Northern Territory, Darwin, 7 July.

National Archives of Australia 2003, *Cyclone Tracy, Darwin*, Fact sheet 176, 22 May.

Navy News 1975, vol. 18, no. 1, January 17, 1975.

Navy News 1975, vol. 18, no. 2, January 31, 1975.

Navy News 1975, vol. 18, no. 3, February 14, 1975.

Office of the Flag Officer Commanding, East Australia Area, Garden Island 1975, *Report of the assistance and aid given by the Command of Flag Officer Commanding, East Australia Area, to the people of Darwin, the Fleet and other organisations*, N2/2/1863, 20 May.

RAN Administrative Authority: Naval Officer in Charge and Naval Officer Commanding, North Australia 1974, *Reports of Proceedings, HMA Ships and Establishments, AWM78 Class 403*, AWM78 403/5, October-December, Naval Officer Commanding, North Australia.

RAN Administrative Authority: Naval Officer in Charge and Naval Officer Commanding, North Australia, 1975, *Reports of Proceedings, HMA Ships and Establishments, AWM78 Class 403*, AWM78 403/6, January-March, Naval Officer Commanding, North Australia.

Spicer, J., 1964, *Report of Royal Commissioner on loss of HMAS Voyager: Melbourne, 13th August 1964*, Commonwealth Government Publisher, Canberra.

Tolkien, J.R.R., 1954, *The Lord of the Rings, Part 1, The Fellowship of the Ring*, Harper Collins, London.

ACKNOWLEDGEMENTS

My deep thanks go to my dear fellow Wran recruits: Lyn Baker (nee Alderton), Joan Smolinski (nee Czarnecki) and Ruth Clarke (nee Turnbull) (dec). It was an honour and a privilege to serve with you and I treasure our enduring friendship.

My thanks, admiration and respect also goes to all the other people who told me their stories with such candour: Keith Chay, Barry Chester, Captain Robert 'Bob' Dagworthy AM RAN Rtd and Christine Dagworthy, Michael Daly, Commander Roger Fairbairn RAN Rtd, Michael and Rosemary French, Lieutenant Commander Simon Gourlay RAN Legal Officer Rtd, Liz Grimes, Mick Hamilton, Lieutenant Commander David John RAN Rtd, Bernie Kloss, Patricia 'Pat' Nuss, Michael and Christine Pilkington, Commander Les Renfrey RAN Rtd, Commander WG 'Bill' Ritchie AM RAN Rtd (dec), Ron Richards, Vice Admiral David Shackleton AO RAN Rtd, Captain Brian Swan AM RAN Rtd, Carolyn Spencer, Richard Usher OAM and Bronwen Usher, and Commander Tony Vine RAN Rtd.

Melanie Myers of the Queensland Writers Centre assisted me in raising the standard of my writing. Her encouragement helped me through many crises of confidence.

Hachette Australia's publisher Sophie Hamley, in-house editor Annie Zhang, copyeditor Samantha Sainsbury and proofreader Annabel Adair were unfailing in their enthusiasm for this book. They made me feel part of a tight team.

My deepest love and gratitude are for my son Terry Collins and my daughter-in-law Amanda Ford Collins. I endured many long dark nights of the soul while writing this book and they were with me every step of the way. This book would never have seen the light of day without their steadfast support.

INDEX